Cutting-edge Research on the Action Plan for Prevention and Control of Emerging Contaminants in China

In Focus – Special Book Series

Cutting-edge Research on the Action Plan for Prevention and Control of Emerging Contaminants in China

Editors

Ye Du, Hai-Xiang Li, Yun Lu, Dong-Bin Wei and Wen-Long Wang

Published by   **IWA Publishing**
Unit 104-105, Export Building
1 Clove Crescent
London E14 2BA, UK
Telephone: +44 (0)20 7654 5500
Fax: +44 (0)20 7654 5555
Email: publications@iwap.co.uk
Web: www.iwapublishing.com

First published 2023
© 2023 IWA Publishing

British Library Cataloguing in Publication Data
A CIP catalogue record for this book is available from the British Library

ISBN: 9781789064803

Contents

doi: 10.2166/wrd.2023.001

Editorial: Cutting-edge research on the action plan for the prevention and control of emerging contaminants in China

Water treatment and reuse is the key strategy to address the problems of water shortages. To facilitate the communication and application of technology and research in water reclamation, the National Conference on Water Treatment and Reuse of China has been held annually since 2017 by the Committee of Water Treatment and Reuse, Chinese Society for Environmental Sciences.

The theme of the 7th conference was the Water Recycling and Integrated Management of Water Systems. The conference, held from April 14 to 16 2023, brought together researchers working in different areas related to water treatment and reuse. More than 13,000 researchers from universities, institutes, and companies attended the conference.

The conference covered 20 topics:

1. Physical separation theory and technologies
2. Chemical transformation theory and technologies
3. Biological purification theory and technologies
4. Natural treatment theory and technologies
5. Water treatment theory and technology based on microalgae
6. Chemical risk and control
7. Disinfection and microbial risk control
8. Standard and risk control of reclaimed water environment reuse
9. Reclaimed water storage and reuse
10. Water quality detection and water characteristic research methods
11. Standards and policy for water quality control and water resource management
12. New materials, reagents, and equipment for water treatment
13. Design and intelligent operation of water treatment process
14. Urban water system and water recycling
15. Industrial water system and water recycling
16. Rural water system and water recycling
17. Pollution control and utilization of rainwater runoff
18. Water environment constructing and regional water recycling
19. Watershed environment remediation and water recycling
20. Desalination technology and processes

This book brings together seven papers that focus on the prevention and control of emerging contaminants, selected from papers presented at the 7th National Conference on Water Treatment and Reuse of China, and is sponsored by CSCEC Scimee Sci. &Tech. Co., Ltd.

Each of these seven papers focuses on a certain link of water reclamation and reuse. The first four articles focus on water treatment and risk control. Dong *et al.* (2023) investigated the performance and mechanism of sulfonamide antibiotic degradation by plants in karst areas. Liang *et al.* (2023) also focused on antibiotics and presented a study on the removal of antibiotics by graphene oxide membranes. Yao *et al.* (2023) studied the treatment performance and mechanism of gas field wastewater by Fenton/metal sulfide processes. Chen *et al.* (2023) investigated the biofilm yield and biomass distribution in the process of municipal wastewater treatment by a multi-stage anoxic/oxic-moving bed biofilm reactor system.

Furthermore, two articles are related to the evaluation of biological effects of pollutants. Yuan *et al.* (2023) reported the effects of chronic zinc exposure on the oxidative stress and neurotoxicity of zebrafish. Li *et al.* (2023) examined the growth tolerance, bioremoval efficacy, and metabolomic profiles of the cellular responses of *Chlorella pyrenoidosa* to phenol and 4-fluorophenol.

Lastly, one review paper provides new insights for the application of fluorescence technology. Jing *et al.* (2023) summarized recent applications of fluorescence analysis in water research for the characterization of pollutants, evaluation of water treatment processes, and monitoring of emerging contaminants.

This special issue deals with various problems and investigates different treatment technologies concerning emerging contaminants. In response to the challenges of severe water shortages and the deterioration of water quality, the focus of cutting-edge technology and systems for water treatment and reuse is fast-growing worldwide. Progress is needed to provide high-quality reclaimed water, minimize the risks to the environment and human health, and support the use of non-conventional water resources.

Guest Editors
Dr Ye Du
College of Architecture & Environment, Sichuan University, China.

Dr Hai-Xiang Li
Guilin University of Technology, China.

Dr Yun Lu
School of Environment, Tsinghua University, China.

Dr Dong-Bin Wei IWA
Chinese Academy of Sciences, China.

Dr Wen-Long Wang
Tsinghua Shenzhen International Graduate School, China.

REFERENCES

Chen, N., Wang, X., Huang, M., Maletskyi, Z., Ratnaweera, H. & Bi, X. 2023 Quantitative study of biofilm yield and biomass distribution of a multi-stage moving-bed biofilm system. *Water Reuse* **13** (1), 250–268.

Dong, K., Wang, W., Li, M., Zhou, X., Huang, Y., Zhou, G., Xu, Y., Wang, D. & Li, H.-X. 2023 Degradation of sulfonamide antibiotics in the rhizosphere of two dominant plants in Huixian karst wetland, Guangxi, China. *Water Reuse* **13** (1), 18–32.

Jing, Z.-B., Wang, W.-L., Nong, Y.-J., Zhu, P., Lu, Y. & Wu, Q.-Y. 2023 Fluorescence analysis for water characterization: measurement processes, influencing factors, and data analysis. *Water Reuse* **13** (1), 33–50.

Li, M., Ma, L., An, Y., Wei, D., Ma, H., Zhu, J. & Wang, C. 2023 Bioremoval efficiency and metabolomic profiles of cellular responses of *Chlorella pyrenoidosa* to phenol and 4-fluorophenol. *Water Reuse* **13** (1), 97–106.

Liang, Z., Zhao, X., Huang, W., Qi, H. & Wang, C. 2023 Removal of antibiotics with different charges in water by graphene oxide membranes. *Water Reuse* **13** (1), 220–232.

Yao, B., Chen, Y., Wang, M. & Liu, M. 2023 Fenton process enhanced by metal sulfide for treating the actual evaporated mother liquid of gas field wastewater. *Water Reuse* **13** (1), 193–204.

Yuan, Z., Li, R., Li, S., Qiu, D., Li, G., Wang, C., Ni, J., Sun, Y. & Hu, H. 2023 Oxidative stress, neurotoxicity, and intestinal microbial regulation after a chronic zinc exposure: an experimental study on adult zebrafish (*Danio rerio*). *Water Reuse* **13** (1), 82–96.

doi: 10.2166/wrd.2023.009

Quantitative study of biofilm yield and biomass distribution of a multi-stage moving-bed biofilm system

Ning Chen[a], Xiaodong Wang [a,*], Mei Huang[a], Zakhar Maletskyi[b], Harsha Ratnaweera[b] and Xuejun Bi[a]

[a] School of Environmental and Municipal Engineering, Qingdao University of Technology, Fushun Road 11, Qingdao 266033, China
[b] Faculty of Science and Technology, Norwegian University of Life Sciences, P.O. Box 5003, 1432 Aas, Oslo, Norway
*Corresponding author. E-mail: wangxiaodong@qut.edu.cn

XW, 0000-0001-6333-3740

ABSTRACT

A multi-stage anoxic/oxic (A/O) moving-bed biofilm reactor (MBBR) system with multiple chambers was established for municipal wastewater treatment. The active biomass quantity, bioactivity, and biomass yield of a pilot-scale multi-stage MBBR were investigated in this study. The microbial activity and heterotrophic yield coefficients (Y_H) were measured using respirometric techniques in each chamber at different temperature conditions. Meanwhile, the growth, nitrification, and denitrification rates of functional biomass were also quantified as specific respiration rate (SOUR). The total active biomass in the multi-stage A/O-MBBR system was 0.71–1.68 g COD/m^2 for the aerobic reactor and 0.39–1.44 g COD/m^2 for the anoxic reactor at 10–19 °C. The Y_H values for the anoxic reactors were 0.61–0.69, which were comparable to the recommended value of the activated sludge model (ASM1). The correlation coefficient between *Nitrospira* and the autotrophic specific respiration rate (SOUR$_A$) was 0.82. Meanwhile, denitrifying genera showed a significant correlation with the heterotrophic specific respiration rate (SOUR$_H$) and the active heterotrophic biomass (X_H). This study provided insights into biomass distribution and the corresponding kinetic parameters for the multi-stage MBBR systems, which may serve as a reference for process design and trouble shooting.

Key words: active biomass distribution, low temperature, moving-bed biofilm reactor, respirometric technique

HIGHLIGHTS

- The active biomass was quantitatively investigated in a multi-stage MBBR.
- Microbial activity and heterotrophic yield coefficients were quantified in each reactor.
- The use of respirometric techniques to assess microbial performance is feasible.
- The correlation analysis of microbial community with active biomass and SOUR was carried out.
- Data and facts were provided via long-term comprehensive investigation.

GRAPHICAL ABSTRACT

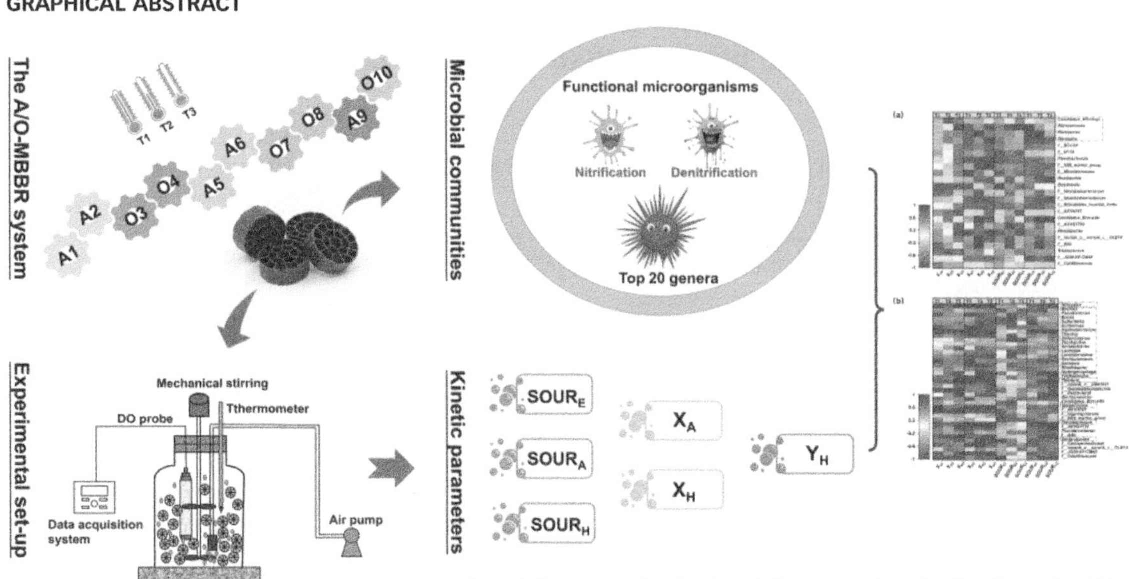

1. INTRODUCTION

Moving-bed biofilm reactors (MBBRs) consist of free-suspended carrier elements, the surface of which provides attachment sites for biofilm growth (Leyva-Díaz *et al.* 2013). The compact design, simple operation, no requirement for sludge recycling, high microbial activity, and stable nitrification make MBBRs highly effective for nitrogen removal from wastewater (Morgan-Sagastume 2018; Ooi *et al.* 2018; Du *et al.* 2022). The MBBR process has a higher biomass residence time compared to the activated sludge method (Avcioglu *et al.* 1998), which provides favorable conditions for the enrichment of nitrifying bacteria with low specific growth rates (Gu *et al.* 2014; Li *et al.* 2020) while ensuring the biomass number. According to the main mode of microbial presence, MBBR systems can be divided into hybrid MBBR (i.e., IFAS) or pure MBBR technology. Due to China's wastewater treatment issues, such as small land area, high standards, and poor stability, IFAS has been successfully used in upgrading Chinese wastewater treatment plants with its unique advantages. However, there are fewer studies investigating the microbial growth and application of pure MBBR, especially for the multi-stage pure MBBR process.

Based on the reaction kinetics principle to optimize the removal process of pollutants, staging biological reactors has been proposed (Joss *et al.* 2006). Due to the advantages of the original MBBR process, the multi-stage MBBR process has become a novel process preference in recent years due to its ability to control the concentration and nature of the growth substrate in contact with the biomass and the ability to grow different types of active microbial systems in a targeted manner (Polesel *et al.* 2017; Torresi *et al.* 2017). Additionally, it was shown that significant differences in microbial communities for nitrogen removal and pollutants biotransformation in the multi-stage MBBR were determined by the different gradients of organic loading and mass concentrations (Polesel *et al.* 2017). Furthermore, the special operating mode of multiple reactors in series is more conducive to screening and enrichment of dominant nitrifying bacteria (Zhang *et al.* 2019; Fan *et al.* 2022). At present, the research on the multi-stage pure membrane A/O-MBBR system mostly focuses on the conversion relationship of carbon and nitrogen removal, while the complex biomass distribution and activity that play a key role in this relationship are less studied.

The biological treatment process of wastewater is completed through a series of complex microbial metabolic processes. Low temperature can directly affect the growth and metabolism of biomass and inhibit microbial activity, especially the inhibition of autotrophic nitrifying bacteria (Antoniou *et al.* 1990). Although MBBR is considered an option to ensure the effective removal of pollutants under low-temperature conditions, the weakening effect of low temperature on pollutant removal efficiency during winter is not negligible.

Delatolla *et al.* (2010) discovered that when the temperature dropped from 14 to 3 °C, the ammonium nitrogen removal rate decreased by 73% (from 0.60 to 0.16 kg-N/(m^3·d)). Ashkanani *et al.* (2019) noticed a reduction of 54.7, 62.3, and 89%, respectively, in the surface ammonia removal rate of MBBR systems using different bio-carriers as the temperature decreased from 35 to 4 °C. Regarding the effect of temperature on denitrification, it has been shown that low temperatures directly affect the hydrolysis rate of substrates involved in denitrification and the activity of denitrifying reductases (Fu *et al.* 2022). While denitrification will be abolished at temperatures below 3 °C (Ghafari *et al.* 2008), Young *et al.* (2017a) showed that a temperature dropping to 5 °C led to a decrease in denitrification rates by 50%. Whereas studies reflecting the effect of temperature on nitrification and denitrification in terms of ammonia removal rates are adequate, it remains to be considered how temperature directly affects biomass distribution and activity which influences nitrification and denitrification from a microbiological perspective. Young *et al.* (2017a) showed that the increase in biofilm thickness of the MBBR system at low temperatures can compensate for the increase in the ammonia removal efficiency of single cells to improve the nitrification performance. Therefore, the number of active biomass and the cells' activity per unit surface area largely determine the final treatment effect of the system. Studies investigating the inhibition effect of low-temperature conditions focused on the influence on nitrification efficiency, while the effect of low temperature on heterotrophic denitrification remains to be elucidated. Moreover, it has been previously shown that the microbial activity of pure MBBR occurs only in the biofilm, even when the wastewater contains exfoliated biofilms and free-growing microbial suspension carriers (Piculell *et al.* 2014; di Biase *et al.* 2021). In that case, it is particularly important to consider biomass composition, activity, and diversity when assessing its function (Helbling *et al.* 2015; Johnson *et al.* 2015; Torresi *et al.* 2018).

Therefore, the use of respirometry combined with the Activated Sludge Model No 1 (AMS1) equation (Henze *et al.* 2000; Ferrai *et al.* 2010), which consists of biofilm growth and decay, would represent a tool to measure microbial activity and composition to obtain the number and kinetic parameters of active biomass in the reactor while reflecting the activity of different types of biomass (Bina *et al.* 2018). The oxygen uptake rate (OUR, mg COD $\cdot$ (min $\cdot$ m^2)$^{-1}$) is an important parameter in respirometric techniques for assessing biomass viability. Different specific respiration rates (SOUR, mg COD $\cdot$ (g $\cdot$ min)$^{-1}$) indicate the respiration rate per unit mass of biofilm. Current respirometry techniques have been used to assess various aspects such as the kinetics of bacteria and activated sludge (Sánchez-Zurano *et al.* 2022); the biodegradability of municipal and industrial wastewater (Hayet *et al.* 2016); biomass activity and wastewater characterization in UCT-MBR pilot plants (Di Trapani *et al.* 2011); the toxicity of pharmaceutical compounds in activated sludge (Vasiliadou *et al.* 2018); and the effect of alkyl phenolic compounds on kinetic coefficients and biomass activity in MBBR (Bina *et al.* 2018). Respirometry has become a common technique for assessing microbial activity and characterizing microbial performance. Considering that the type of substrate in the reactor affects the kinetic parameters (Amin *et al.* 2013; Mansouri 2014), it is necessary and beneficial to determine the kinetic parameters of the more sensitive microorganisms under different conditions to develop a suitable design for simulating biological processes. Among them, the accurate quantification of the yield coefficient of heterotrophic biomass (Y_H) will promote the reasonable control of source dosage under low-temperature conditions in winter.

Here, we evaluated the effect of temperature changes on the distribution and activity of active biomass in a multi-stage MBBR system. With the help of respirometric techniques and the ASM1 equation, we characterized biomass viability. A method for reflecting the nitrification and denitrification capacity per unit mass of biomass in terms of SOUR was described and the effect of temperature variation on this as well as on Y_H was reported. High-throughput techniques aid in analyzing the correlation between microbial communities and active biomass. We provided a simple and rapid method to measure and compare the microbial performance in each reactor of the multi-stage MBBR system.

2. MATERIALS AND METHODS

2.1. Pilot MBBR system description

Samples for this study were obtained from a pilot system at a municipal wastewater treatment plant in Qingdao, China, between November and December 2021. The system was in a long-term stable operation state before the test. A schematic representation three-stage A/O-MBBR pure membrane pilot system is shown in Figure 1. The overall length × width × height of the pilot system was 14 m × 1 m × 1.7 m, divided equally into 10 reactors, which consisted of three-stage sub-systems in series. The first A/O subsystem consisted of two anoxic reactors (A1 and A2) and two oxic reactors (O3 and O4,), the second A/O subsystem consisted of two anoxic reactors (A5 and A6) and two oxic reactors (O7 and O8,), while the third A/O subsystem consisted of one anoxic reactor (A9) and one oxic reactor (O10). Each reactor was filled with cylindrical-shaped carriers made of polyethylene with a volumetric filling rate of 50%. Each carrier has a diameter of 25 mm, a height of 10 mm, and a specific surface area of 500 m^2/m^3. The system adopted a two-point continuous water inlet (ratio

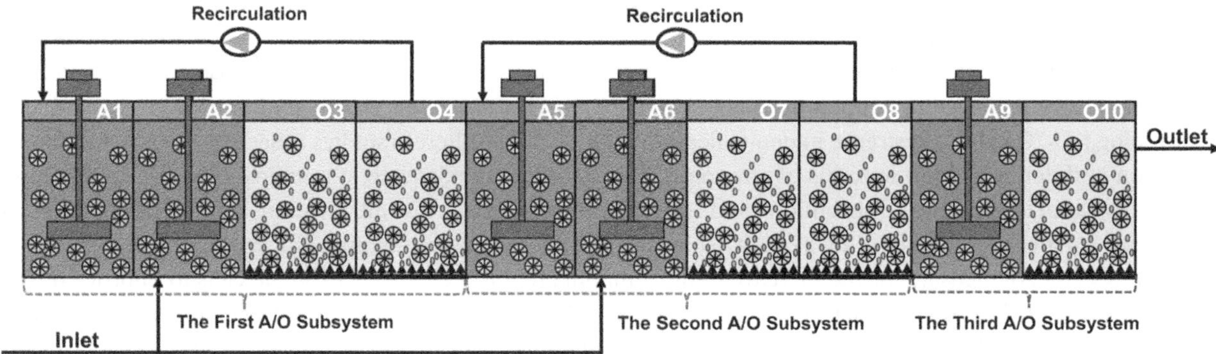

Figure 1 | Schematic representation of the three-stage A/O-MBBR pilot system.

of inlet flow: 1:1), with water inlet points at A2 and A6, and a total water inlet flow of 28.8 m^3/d. The cross-sectional flow rate between each reactor was 1.2 m/h. The total hydraulic retention time was 11.7 h. Nitrate recycling was set up in two stages, the first stage was set up from the O4-MBBR to the A1-MBBR (sludge return ratio 200%) and the second stage was set up from O8-MBBR to A5-MBBR (sludge return ratio 100%). The stirrer in the anoxic reactor had a speed of 90–110 r/min and the dissolved oxygen (DO) in the aerobic reactor was maintained at 3.5–9.6 mg O_2 /L by aeration, by which the adequate fluidization of the bio-carriers was ensured. To ensure smooth post-denitrification, 50 mg COD/L sodium acetate was added to the A9-MBBR as the exogenous carbon source. The samples were taken in winter with temperatures ranging from 10 to 19 °C. Divided into three temperature stages according to the variation of the system temperature with the seasons (T1: 16–19 °C, T2: 13–16 °C, T3: 10–13 °C).

2.2. Wastewater characteristics

The system inlet water was pumped from the effluent primary sedimentation tank of the wastewater treatment plant, and an air floatation tank was applied as additional pretreatment for particle separation before the raw wastewater entered the pilot system (inlet wastewater characteristics are shown in Table 1). The water quality parameters for each reactor stage are shown in Table 2. Due to variation of the actual sewage, the above water quality parameters are within the normal range of variation.

2.3. Respirometric tests

The respiration measurement device for this experiment is based on the continuous measurement of DO in a closed apparatus, with subsequent calculation of the OUR of autotrophic and heterotrophic bacteria. As shown in Figure 2(a), the apparatus consists of a cylindrical Plexiglas reactor with an effective volume of 5 L. Heating tape is fixed on the outside of the apparatus to ensure that the temperature during the test is consistent with the temperature at the time of sampling.

The good fluidization state of the suspended bio-carrier was ensured by a wavemaker, which allows full contact with the nutrients in the water. An aeration system with a 6 cm bread bubble tray was connected to an 8 L/min aeration pump via an

Table 1 | Experimental conditions and wastewater quality characteristics of the wastewater fed to the multi-stage A/O-MBBR system

Stage	Reactor	DO (mg/L)	SCOD (mg/L)	NH$_4^+$-N (mg/L)	NO$_2^-$-N (mg/L)	NO$_3^-$-N (mg/L)
Stage 1	A1	0.86 ± 0.16	42.4 ± 17.7	4.9 ± 1.8	0.62 ± 0.25	10.21 ± 1.9
	A2	0.2 ± 0.03	59.45 ± 12.0	16.8 ± 1.3	0.8 ± 0.22	2.53 ± 1.40
	O3	4.77 ± 0.35	44.00 ± 14.1	12.4 ± 1.8	1.48 ± 0.28	4.17 ± 1.58
	O4	3.23 ± 0.73	39.7 ± 9.7	5.6 ± 1.8	1.49 ± 0.47	9.09 ± 1.93
Stage 2	A5	0.9 ± 0.18	51.2 ± 15.6	4.65 ± 1.7	1.15 ± 0.51	9.72 ± 1.82
	A6	0.18 ± 0.03	60.5 ± 6.4	18.7 ± 2.6	0.92 ± 0.17	1.80 ± 1.28
	O7	4.19 ± 0.25	46.8 ± 8.8	12.05 ± 2.3	1.64 ± 0.39	4.68 ± 1.73
	O8	5.09 ± 0.15	50.75 ± 17.55	6.2 ± 1.8	1.97 ± 0.93	9.25 ± 1.97
Stage 3	A9	0.3 ± 0.04	47.5 ± 7.5	5.25 ± 1.4	3.74 ± 2.62	1.25 ± 0.66
	O10	5.08 ± 0.63	35.1 ± 24.2	1.85 ± 1.6	1.07 ± 0.67	6.35 ± 2.26

Table 2 | The water quality parameters of the multi-stage A/O-MBBR system in each reactor

Inlet flow (m³/d)	Nitrate recycling ratio	SCOD (mg/L)	NH₄⁺-N (mg/L)	TIN (mg/L)	Temperature (°C)	pH
28.8	200% and 100%	141.46 ± 24.31	39.50 ± 4.94	40.60 ± 5.00	10–19	7–9

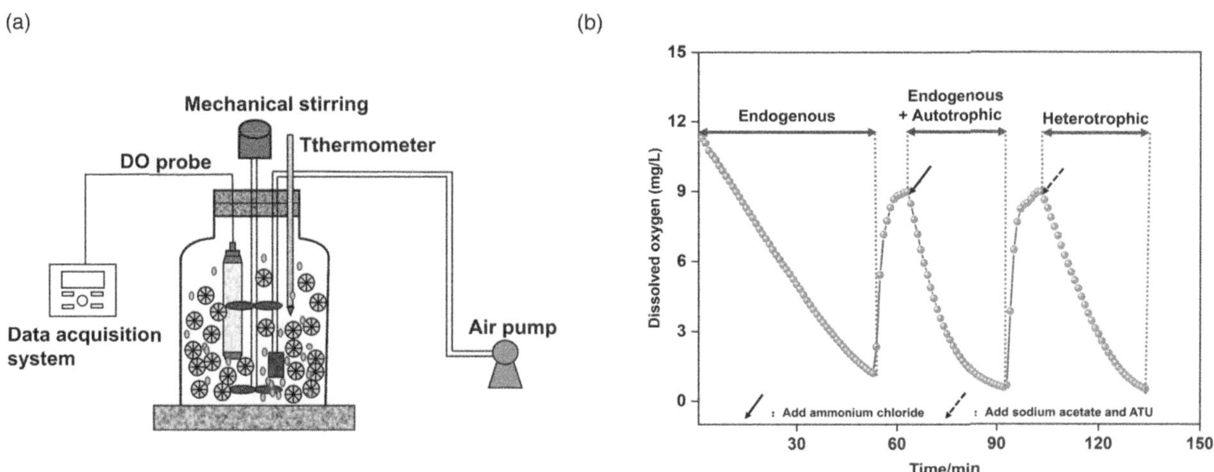

Figure 2 | The batch set-up for respiration tests. (a) The SOUR measurement system and (b) the typical DO profile with endogenous, autotrophic, and heterotrophic respiration rates of the biofilm.

oxygen pipe and operating at a constant airflow. A portable DO meter, the Hash HQ40d (HACH, Loveland, CO, USA), was used in connection with the data collection system for real-time monitoring of the DO concentration in the reactor at a frequency of 1 min. A wireless temperature measurement device was used to obtain the reactor's internal temperature in real-time to facilitate the timely treatment of temperature abnormalities.

2.4. Batch respirometric experiments

Respirometry is commonly used for the control of activated sludge processes by estimating the ASM1 parameters (Avcioglu *et al.* 1998; Collivignarelli *et al.* 2019). Based on the aerobic biomass respiration, OURs of the suspended bio-carriers were assessed. The process was divided into three stages to obtain the endogenous respiration rate (OUR$_E$), the autotrophic respiration rate (OUR$_A$), the heterotrophic respiration rate (OUR$_H$), the active autotrophic biomass (X_A), the active heterotrophic biomass (X_H) and Y_H, where the specific steps and analyses for the quantification of X_A and X_H were carried out according to Wang *et al.* (2018). The equations of X_A, X_H, and SOUR are shown as Equations (1)–(3), where the definitions of notions are listed in Table 3.

$$X_A = \frac{1}{\mu_{A,Max}} \cdot \frac{Y_A}{1 - Y_A} \cdot \text{OUR}_A \tag{1}$$

$$X_H = \frac{1}{\mu_{H,Max}} \cdot \frac{Y_H}{1 - Y_H} \cdot \text{OUR}_H \tag{2}$$

$$\text{SOUR} = \frac{\text{OUR}}{\text{MLSS}} \tag{3}$$

where the unit of X_A and X_H was g (COD)/m², the m² was relative to the surface area of biomass. The unit of specific biomass quantity is converted according to Equation (4).

$$X(g(\text{COD})/m^2) = \frac{X_1(mg(\text{COD})/L) * V}{N * S * 1,000} \tag{4}$$

Table 3 | Variables and model parameters of biomass growth and respiration rate models

Notation	Definition	Units
X_A	Active autotrophic biomass	g COD/m^2
X_H	Active heterotrophic biomass	g COD/m^2
Y_A	Yield for autotrophic biomass	g COD $\cdot$ (g N)$^{-1}$
Y_H	Yield for heterotrophic biomass	g COD $\cdot$ (g COD)$^{-1}$
$\mu_{A,max}$	Maximum specific growth rate for autotrophic biomass	day^{-1}
$\mu_{H,max}$	Maximum specific growth rate for heterotrophic biomass	day^{-1}
OUR	Respiration rate	mg COD $\cdot$ (min $\cdot$ m^2)$^{-1}$
OUR$_A$	Autotrophic respiration rate	mg COD $\cdot$ (min $\cdot$ m^2)$^{-1}$
OUR$_H$	Heterotrophic respiration rate	mg COD $\cdot$ (min $\cdot$ m^2)$^{-1}$
SOUR	Specific respiration rate	mg COD $\cdot$ (g $\cdot$ min)$^{-1}$
MLSS	Mixed liquor suspended solids	g/m^2

where V was the effective volume of the reactor (L), N was the number of bio-carriers, S was the specific surface area of the one bio-carrier (m^2).

During the first step, the suspended bio-carriers were rinsed 3–5 times with tap water to remove the attached organic matter, allowing them to enter the endogenous respiration phase, which can be verified by continuous monitoring of the DO curves. Full aeration was carried out to make the DO concentration in the reactor reach the saturated DO (10–12 mg O$_2$/L) before entering the next stage. During the second stage, when the DO concentration in the batch set-up reached 10–12 mg O$_2$/L, a prepared ammonium chloride solution of a certain concentration was measured with a pipette and injected into the reaction device to ensure the initial concentration of 20 mg N/L. While during the third stage, 20 mg N/L allylthiourea was added to inhibit nitrification, and sodium acetate at a dose of 250 mg COD/L was used for heterotrophic biomass respiration. The batch set-up was sealed at all stages to avoid interference from the outside air and the DO profiles were monitored. The DO profiles for the three stages of the test procedure are shown in Figure 2(b).

2.5. Yield coefficient of heterotrophic biomass

The batch method of activated sludge and the respirometric method was commonly used to estimate the Y_H (Strotmann *et al.* 1999; Vanrolleghem *et al.* 1999; Gujer 2006). In this study, we made improvements based on Ochoa *et al.* (2002), Wang *et al.* (2018), and some other studies, and successfully applied the batch respirometric method. Considering the possibility of biomass denitrification in the anoxic under low DO (DO < 2 mg/L) conditions (Fathali *et al.* 2019), the Y_H estimation was revised. According to the mechanism of organic matter consumption through heterotrophic biomass aerobic respiration, the equation of Y_H value is shown as Equation (5).

$$Y_H = \frac{\Delta COD - \int \Delta OUR(t)dt - \Delta NO_3^-}{\Delta COD} \tag{5}$$

where ΔCOD was the difference in COD concentration between the beginning and end of the third stage (mg/L), $\int \Delta OUR(t)dt$ was the integral value of OUR at any time during the third stage, ΔNO_3^- was the difference in COD concentration between the beginning and end of the third stage (mg/L).

2.6. High-throughput sequencing methods

During the operating period, eight carriers were randomly selected from each reactor. Subsequently, biofilms were immediately scraped from the carriers and stored at −80 °C. Samples were named A1_1, A1_2, and A1_3 (A1_1, A1_2, and A1_3 represent the A1 reactor at T1, T2, and T3 respectively). Both microbial community genomic DNA extraction and DNA sequencing were performed according to the standard protocols by Majorbio Bio-Pharm Technology Co., Ltd (Shanghai, China) and the detailed steps were previously described (Zhang *et al.* 2020). The hypervariable region V3–V4 of bacterial

16S rRNA gene was amplified with the primer pairs 338F (5′-ACTCCTACGGGAGGCAGCAG-3′) and 806R (5′-GGAC-TACHVGGGTWTCTAAT-3′).

2.7. Statistical methods and software

The correlation analysis in this study was performed on origin 2021 Pro using the Principal Component Analysis plug-in, and the experimental data were statistically analyzed and pre-processed using Microsoft Excel 2019.

3. RESULTS AND DISCUSSION

3.1. Active biomass

3.1.1. Overall distribution of active biomass

Multi-stage MBBR systems have different concentrations of substrates within each reactor due to their installation conditions. Biofilms in different reactors are exposed to different ratios of substrate types and different available conditions (e.g., readily biodegradable organics, slowly biodegradable organics, or hard-to-biodegrade organics). The pilot system is divided into each reactor, which acts alone as a specific functional area to remove conventional pollutants (e.g., organic carbon and nitrogen), there are differences in the external environment for microbial growth, which can affect the growth of microbial types in a targeted manner. The environmental conditions combined with substrate conditions determine the number of active biomass and the proportion of microbial types in each reactor of a multi-stage MBBR system (Revilla *et al.* 2016). As shown in Figure 3, the total active biomass was the highest in the O3-MBBR and O7-MBBR with 1.6835 and 1.5512 g COD/m^2, respectively. As the inlet points are located at A2-MBBR and A6-MBBR, the substrate content in A2-MBBR and A6-MBBR with their

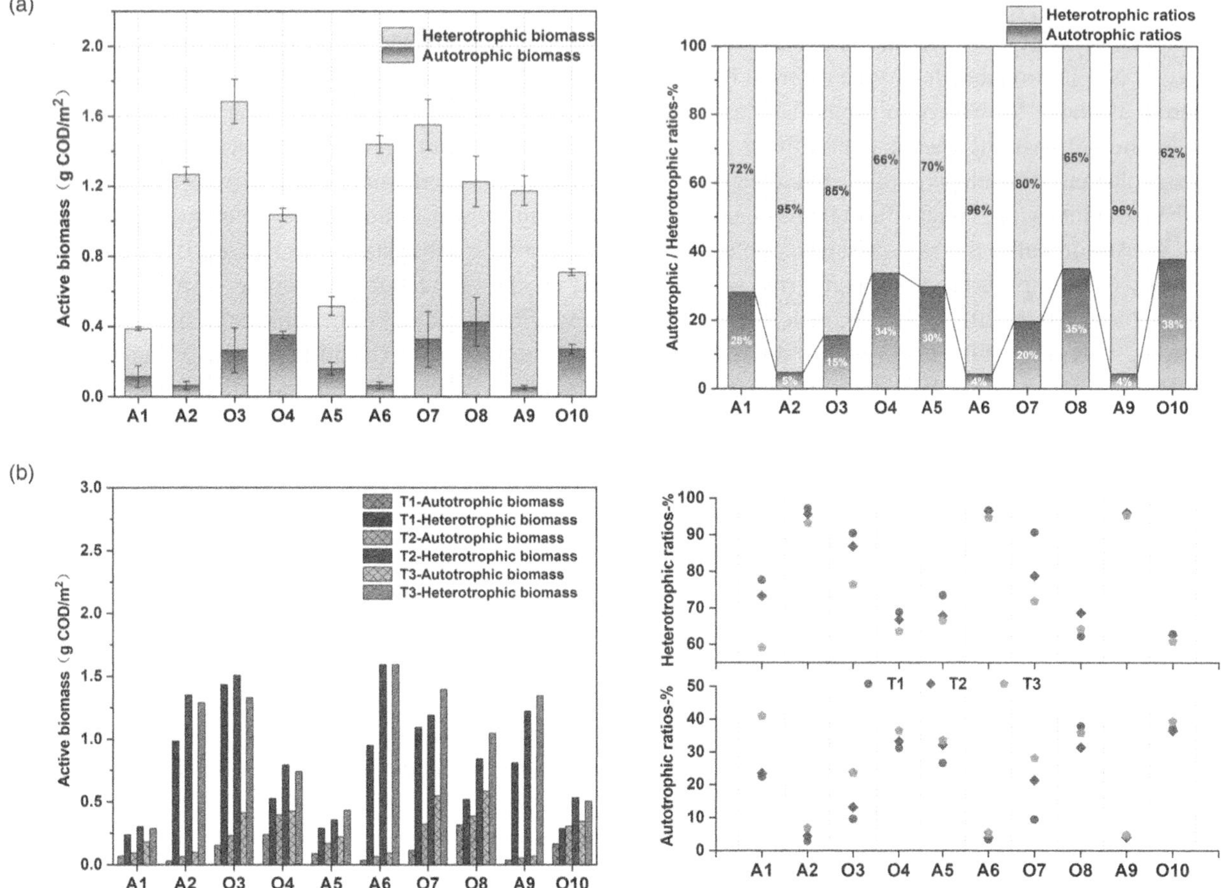

Figure 3 | Active biomass distribution of the biofilm at each sub-reactor. (a) Heterotrophic and autotrophic biomass quantity and proportion and (b) variation of the biofilm active biomass distribution at different temperatures.

subsequent O3-MBBR and O4-MBBR was relatively adequate and more favorable to microbial growth, which is also the reason why A2-MBBR and A6-MBBR (A2: 1.2670 g COD/m^2; A6: 1.4381 g COD/m^2) had a much higher total active biomass than that of A1-MBBR and A5-MBBR (A1: 0.3870 g COD/m^2; A5: 0.5154 g COD/m^2). Furthermore, the A5-MBBR and A6-MBBR reactors receive substrates from the first MBBR reactor, where the difficult biological substrates may have been transformed due to longer residence time and aeration in the O3-MBBR and O4-MBBR and can be used by the biomass, which may also account for the slightly higher total active biomass of A5-MBBR compared with A1-MBBR, and the same as A6-MBBR compared with A2-MBBR. The O3-MBBR and O7-MBBR environments were more favorable for autotrophic biomass growth due to external environmental regulation (e.g., aeration conditions, low C/N ratio conditions), which in turn increased the overall active biomass. Along with a further reduction in substrate loading and creating an environment with a low C/N ratio, the growth of autotrophic biomass was in a favorable position (Kumar *et al.* 2012; Sun *et al.* 2014). Although the total active biomass was lower for O4-MBBR than O3-MBBR and lower for O8-MBBR than O7-MBBR (O4: 1.0353 g COD/m^2, O8: 1.2262 g COD/m^2), the proportion of autotrophic biomass was higher for O4-MBBR than O3-MBBR and higher for O8-MBBR than O7-MBBR (O4: 34% > O3: 15%; O8: 35% > O7: 20%) (Figure 3(a)). Additionally, the active autotrophic biomass of O4-MBBR and O8-MBBR were higher than the second aerobic reactor reported by Wang *et al.* (2018) (21.2%), 14.7% for the intermittently aerated MBBR system reported by Luan *et al.* (2022), and the activated sludge reported by Fernandes *et al.* (2014) (12.2%). A previous study has shown that only 0.5–4% of the activated autotrophic biomass was present in a full-scale activated sludge system (Ma *et al.* 2015). Moreover, the influence of the nitrate recycling carrying large amounts of DO created conditions for autotrophic biomass growth; thus, leading to A1-MBBR and A5-MBBR autotrophic biomass proportions reaching 28 and 30%, respectively, even higher than in the O3-MBBR and O7-MBBR nitrification function reactor. Excessive DO levels have an impact on the growth of dominant populations in the system while competing with denitrifying bacteria for carbon sources, severely limiting the denitrification effect. This was one of the reasons why the system had set the inlet point at A2-MBBR and A6-MBBR while improving the carbon source utilization. This indicates that the reasonable setting of the nitrate recycling point and the nitrate recycling ratio is essential for improving the denitrification effect of the system. A9-MBBR was the anoxic reactor with the highest proportion of heterotrophic biomass in the system (96%), probably due to the external injection of sodium acetate as the readily biodegradable organic matter is more readily available to the heterotrophic biomass (Torresi *et al.* 2017; Pelaz *et al.* 2018). The proportion of autotrophic biomass in the O10-MBBR was at a maximum of 38% (end of the system) and the reactor contained fewer electron donors (organic matter) while most of the biodegradable organic matter has been removed in the front-end system, preventing anaerobic alienation while creating an environment for the dominant growth of autotrophic biomass (Wu *et al.* 2014; He *et al.* 2018). Additionally, the ratio of autotrophic biomass varied as the substrate load changed throughout the system; thus, it was worth considering the specific role of organic load on the growth of autotrophic biomass (Almomani *et al.* 2014; Young *et al.* 2016, 2017a, 2017b), which can provide a reference for us to control the ratio and abundance of autotrophic biomass by controlling the organic load and ultimately monitoring the nitrification activity of the aerobic reactor.

The above analysis showed that the multi-stage A/O-MBBR system was favorable to the formation of different proportions of active biomass and different functional zones met the treatment needs of different pollutants. Furthermore, the proportion of autotrophic biomass in the MBBR system was much higher than that in the activated sludge system (Fernandes *et al.* 2014), which was more favorable to nitrification.

3.1.2. Active biomass distribution at different temperatures

Temperature is one of the most important factors that determines the MBBR performance. While nitrification is particularly sensitive to temperature and stops below 8 °C (Hurse & Connor 1999), MBBR's unique structure can reduce the wear of biofilm and outperforms other biological treatment technologies even in harsh low-temperature environments (Young *et al.* 2017a).

Figure 3(b) illustrates the active biomass variation in each reactor at different temperatures for a multi-stage A/O-MBBR. Overall, it can be seen that the effect of temperature on the active biomass was still relatively large. With the decrease in temperature, the total active biomass showed an upward trend. In particular, a larger increase in total active biomass can be seen when the temperature was reduced from T1 to T2, with an increase of approximately 25–87%. In contrast, the maximum increase in total active biomass was only 24% when the temperature was reduced from T2 to T3; thus indicating that low temperatures limited biomass growth. Interestingly, the total active biomass of each reactor in the first A/O-MBBR subsystem decreased during this process, while MLSS (Supplementary material, Table A1) did not; this indicated that the proportion of active material within the biofilm decreased with lower temperatures, whereas the second A/O-MBBR subsystem was relatively unaffected. The decrease in total active biomass in the first A/O-MBBR subsystem during temperature change is mainly due to the drop in X_H. The maximum growth of X_H in the second A/O-MBBR subsystem was only 24%, a decrease of 62% compared to the maximum growth of X_H from T1 to T3. Furthermore, during the whole temperature change, the X_A in the aerobic reactor maintained a high growth rate, which was more obvious between T1 and T2, increasing by 51–184%, with an increase of about 8–79% from T2 to T3 and a slight decrease in the relative growth, which showed that further decreases in temperature lead X_A growth inhibition. Overall, the X_A growth is more tolerant of low temperatures than X_H and is closely linked to changes in water conditions.

The variation in the respective ratios of X_A and X_H in each reactor at different temperatures for the multi-stage A/O-MBBR is shown in Figure 3(b). Firstly, the X_H ratios in each reactor decreased with dropping temperature, while the corresponding X_A proportion increased. This indicates that the competitive advantage of autotrophic biomass growth was enhanced at low temperatures. Furthermore, we demonstrated that the growth of autotrophic biomass was more tolerant to low temperatures compared to heterotrophic biomass. The most significant increase in ratios of O3-MBBR and O7-MBBR in the aerobic reactor during the temperature decrease (T1 to T3) was from 10 to 24% and from 9 to 28%, respectively, which was equivalent to the decrease in heterotrophic biomass growth within the reactor. However, the increase in X_A ratios was 5 and 4% for O4-MBBR and O10-MBBR, respectively, with a decrease for the O8-MBBR, which may be related to the greater increase in active heterotrophic biomass. Due to the DO from reflux, the X_A ratios in A1-MBBR and A5-MBBR increased by 16 and 11%, respectively, with the nitrate recycling ratio in the first A/O-MBBR subsystem being higher than that in the second A/O-MBBR subsystem. Overall, the active biomass ratio characteristics indicate that autotrophic biomass had a greater growth advantage at low temperature.

3.2. Respiration rate

3.2.1. Endogenous specific respiration rate

As an important parameter in respirometry, the respiration rate is often used to assess the biological capacity of biomass (Bina *et al.* 2018). SOUR refers to the rate of respiration per unit mass of biofilm, and different types of SOURs have different meanings. The value of the endogenous SOUR ($SOUR_E$) can indirectly represent the total number of decaying cells within the biofilm or the rate of decay. The variation in $SOUR_E$ in the anoxic reactor as a function of temperature is shown in Figure 4(a). The $SOUR_E$ of each reactor showed a decreasing trend with dropping temperature, indicating that the total amount of cell decay in the biofilm was reduced and the rate of cell accumulation was accelerated at low temperatures, which is consistent with the previous discussion that the active biomass increased under these conditions. The $SOUR_E$ of A1, A2, A5, and A6 was approximately 0.01 mg COD $\cdot$ (g $\cdot$ min)$^{-1}$ at T1 and fluctuated above and below 0.03 mg COD $\cdot$ (g $\cdot$ min)$^{-1}$ when the temperature was reduced to T3. However, the $SOUR_E$ of A9-MBBR has been generally lower than that of other anoxic reactors. The A9 reactor was supplemented with sodium acetate as an external carbon source, which may have contributed to its relatively low cell decay rate.

The $SOUR_E$ of aerobic reactors is generally higher than that of anoxic reactors (Figure 4(b)), which is in part related to their own large active biomass base. Furthermore, the aerobic reactor $SOUR_E$ exhibited the same response following variation in temperature as the anoxic reactor, suggesting that the low temperature is more favorable to biomass cell accumulation, which to a certain extent compensates for the reduced rate of degradation of organic matter by individual cells (Young *et al.* 2017a). The $SOUR_E$ of O3, O4, O7, and O8 had a maximum value of about 0.07 mg COD $\cdot$ (g $\cdot$ min)$^{-1}$ at T1, with approximately 0.02–0.04 mg COD $\cdot$ (g $\cdot$ min)$^{-1}$ at T3. Furthermore, the $SOUR_E$ of O4 and O8 was less sensitive to the temperature drop than O3-MBBR and O7-MBBR. The $SOUR_E$ of O10-MBBR was less affected by temperature and was always low (approximately 0.02 mg COD $\cdot$ (g $\cdot$ min)$^{-1}$. It is worth considering that A9-MBBR and O10-MBBR, as the third stage of the A/O-MBBR, both had a lower $SOUR_E$ than their counterparts, and the biomass on the biofilm was much less impacted by the substrate load, as the stable material conditions might have promoted microbial growth. The $SOUR_E$ can be used as a parameter to determine the state of microbial cell death or the state of cell accumulation, thus reflecting the system section operational efficacy.

3.2.2. Autotrophic specific respiration rate

The autotrophic SOUR ($SOUR_A$) can indirectly represent nitrification activity. The $SOUR_A$ declined with decreasing temperature (Figure 4(c)), indicating that low temperatures hindered the nitrification activity of autotrophic biomass. The $SOUR_A$ of O4-MBBR, O8-MBBR, and O10-MBBR was significantly higher than that of O-MBBR 3 and O7-MBBR, indicating a

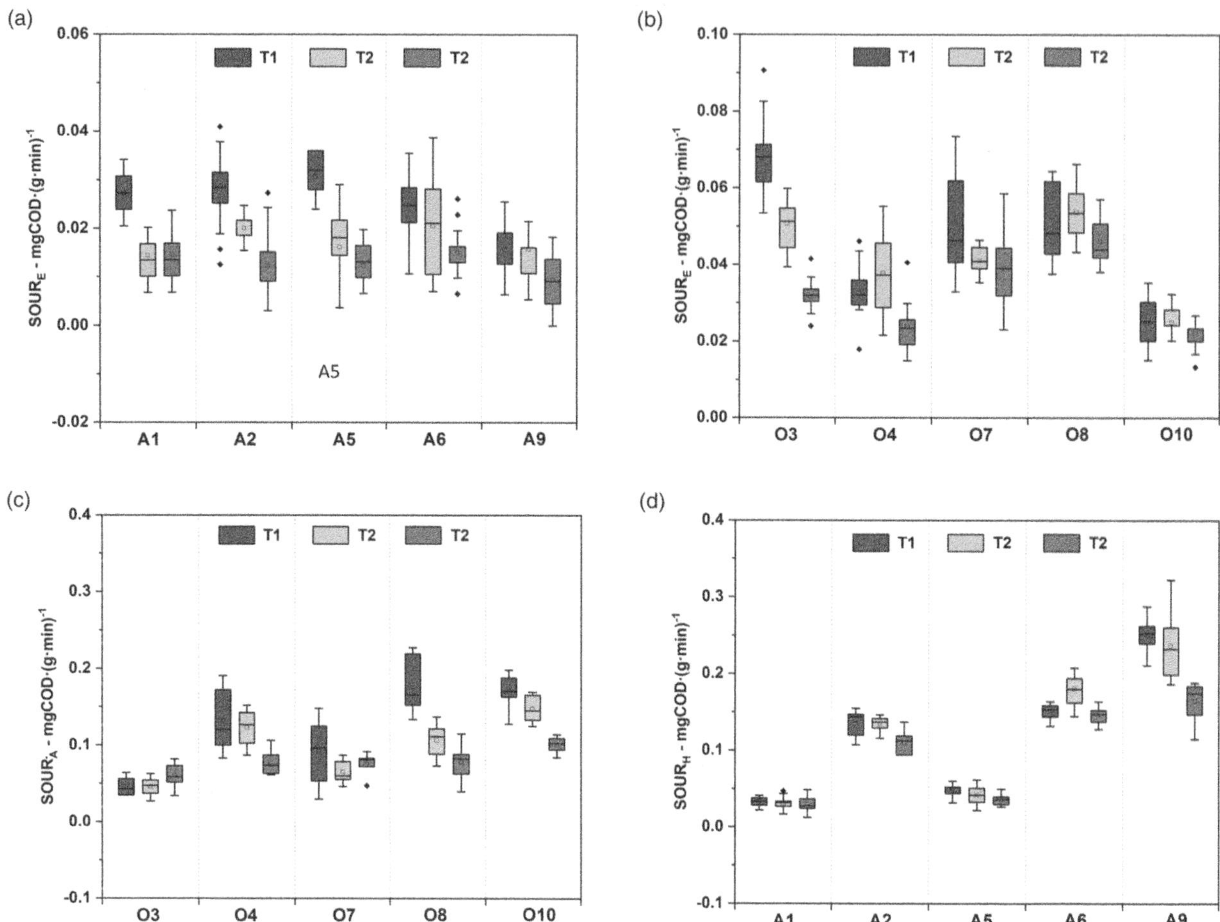

Figure 4 | The SOUR of biofilm at different temperatures. (a) Endogenous SOUR for biofilms in anoxic reactors; (b) endogenous SOUR for biofilms in aerobic reactors; (c) autotrophic SOUR for biofilms in aerobic reactors; and (d) heterotrophic SOUR for anoxic reactors.

relatively good nitrification performance, consistent with their higher X_A. Furthermore, $SOUR_A$ was significantly influenced by temperature, with $SOUR_A$ values for O4-MBBR, O7-MBBR, O8-MBBR, and O10-MBBR varying from 0.1313 to 0.0853, 0.1163 to 0.0748, 0.1817 to 0.1161, and 0.1765 to 0.1083 mg $COD \cdot (g \cdot min)^{-1}$, respectively (Supplementary material, Table A2), as the temperature decreased from T1 to T3, producing a decrease of 35, 36, 36, and 39%, respectively. Additionally, the $SOUR_A$ of O3-MBBR indicates a low nitrification activity, fluctuating around 0.050 mg $COD \cdot (g \cdot min)^{-1}$. Although the total active biomass of the O3-MBBR was consistently high, it was limited by the large proportion of X_H, which led to intense competition between biomass and thus decreased the nitrification activity.

3.2.3. Heterotrophic specific respiration rate

Biological nitrogen removal is carried out through denitrification, which can be divided into autotrophic and heterotrophic denitrification, depending on the carbon source required for biological growth (Karanasios *et al.* 2010). Of these, heterotrophic denitrification has been shown to be the most common and cost-effective process (Wang & Chu 2016). The amount and activity of heterotrophic biomass together determine the effectiveness of biological heterotrophic denitrification. The heterotrophic SOUR ($SOUR_H$) can indirectly represent the denitrification activity per unit mass of biofilm. Temperature had the same effect on $SOUR_H$ as it did on $SOUR_A$ (Figure 4(d)); this suggested that low temperatures have a negative effect on both the nitrification and denitrification activity per unit mass of biofilm. For the five anoxic reactors, $SOUR_H$ was A9 > A6 > A2 > A5 > A1. The substrate of A9-MBBR's microbial contact was superior to that of other anoxic reactors, thus allowing its unit mass biofilm to have a higher active biomass and activity per mass of biofilm and better denitrification, which corresponded to its analytical results for X_H. The $SOUR_H$ values for A1-MBBR, A2-MBBR, A5-MBBR, A6-MBBR, and

A9-MBBR varied from 0.0375 to 0.0274, 0.1439 to 0.111, 0.0535 to 0.0403, 0.1879 to 0.1484, and 0.2395 to 0.1745 mg COD $\cdot$ (g $\cdot$ min)$^{-1}$ (Supplementary material, Table A2) as the temperature decreased from T1 to T3, producing a decrease of 27, 23, 25, 21, and 27%, respectively, which was 10% lower than the decline efficiency of SOUR$_A$. This suggests that the nitrification effect was more heavily influenced by low temperatures than the denitrification (the above is only for a unit mass of biofilm). This it also related to the proportion of active biomass type contained per unit mass of biofilm. The result was in contrast to the positive effect of low temperatures on active biomass growth. Although the increase in active biomass per unit mass of biofilm cannot counteract the effect of low temperature on its activity, the SOUR$_H$ of A6-MBBR was greater at T1 than at T2, which could be due to the large increase in the amount of active heterotrophic biomass during the process (Figure 3(a)), thus offsetting the effect of the weakened activity. The SOUR$_H$ of A1-MBBR and A5-MBBR was significantly lower than in other anoxic reactors. The high redox potential of the nitrifying liquid was generated by the high amount of DO carried by the nitrate recycling, which allowed heterotrophic biomass to use DO for respiratory metabolism and inhibit enzyme activity. Consequently, the anaerobic/anoxic biomass activity was inhibited and can lead to ineffective denitrification (Li *et al.* 2021; Fu *et al.* 2022), which is corroborated by the high proportion of X_A observed.

3.3. Yield coefficient of heterotrophic biomass

The yield coefficient of Y_H is an important parameter in the field of wastewater degradation kinetics and in determining the degradation kinetics of specific chemical substances (Strotmann *et al.* 1999). Supplementary material, Table A3 indicates the Y_H values for each reactor at T3. The higher active heterotrophic biomass at low temperatures were further supported by the fact that more than 80% of the organic matter in the aerobic reactor was used for the growth of heterotrophic biomass during winter low-temperature conditions. As a consequence of nitrate recycling, the A1-MBBR and A5-MBBR were not anoxic in the traditional sense, with approximately 83% of the organic matter playing a role in the material metabolic processes of the heterotrophic biomass. Such high Y_H values were also related to the substrate overdose during the test. Under normal conditions, the biomass in the aerobic reactor was always at low substrate concentrations, and A1-MBBR as well as A2-MBBR were not inlet points with low levels of substrate. Faced with a sudden abundance of organic matter, a large amount was rapidly consumed by heterotrophic biomass for material metabolism, resulting in generally high Y_H values. For the A2-MBBR, A6-MBBR, and A9-MBBR denitrification reactors, which were strictly dominated by heterotrophic biomass, the Y_H values were 0.6068, 0.6231, and 0.6876, respectively, at T3. The Y_H values for A9-MBBR were higher than those for the other two reactors, indicating that the easily biodegradable organic matter was more readily available to the biomass.

The Y_H values of O3-MBBR, O7-MBBR, and A9-MBBR increased with decreasing temperatures (Figure 5), from 0.7956, 0.5352, and 0.5023 (T1) to 0.8867, 0.8604, and 0.6876 (T3), respectively, an increase of 11, 61, and 37%, respectively. The Y_H value of O7-MBBR was most affected by temperature changes, which is consistent with the growth of active heterotrophic biomass of the second stage MBBR system was greater than the first stage through the whole temperature change process. The nitrate recycling ratio of the second stage was half of that of the first stage, and at higher temperatures the DO in the second stage system was not sufficient to be used by both autotrophic and heterotrophic biomass in the aerobic reactor. As the temperature decreased, the oxygen content in the water increased significantly, which was sufficient to satisfy the needs of both types of biomass without competition, thus enhancing the growth space for heterotrophic biomass, which may be one reason why the Y_H value of O7-MBBR was significantly higher than that of O3-MBBR. The Y_H values of A2-MBBR and A6-MBBR were less affected by temperature and decreased with decreasing temperatures, from 0.6718 to 0.6068 and 0.6774 to 0.6231, respectively, with a decrease of 10 and 8%, respectively. This suggests that the ratio coefficient between the growth rate of heterotrophic biomass and the rate of substrate degradation decreased slightly with decreasing temperature, and the consumption of substrate was closely related to the overall number and degradation capacity of heterotrophic biomass. Furthermore, from the perspective of heterotrophic microbial growth only, the carbon source dosing can be appropriately adjusted downwards in winter when the influent load is stable. Moreover, the degree to which denitrification efficiency is affected by low temperatures should be considered when dosing the external carbon sources in practical terms, without wasting energy to maintain a stable system treatment effect.

3.4. Microbial community analysis

The microbial community structure determines the functional properties of a reactor, while temperature and substrate conditions can regulate microbial abundance and diversity at different stages (Wang *et al.* 2019). The microbial community structure of each reactor was evaluated at different temperatures using high-throughput sequencing techniques. The

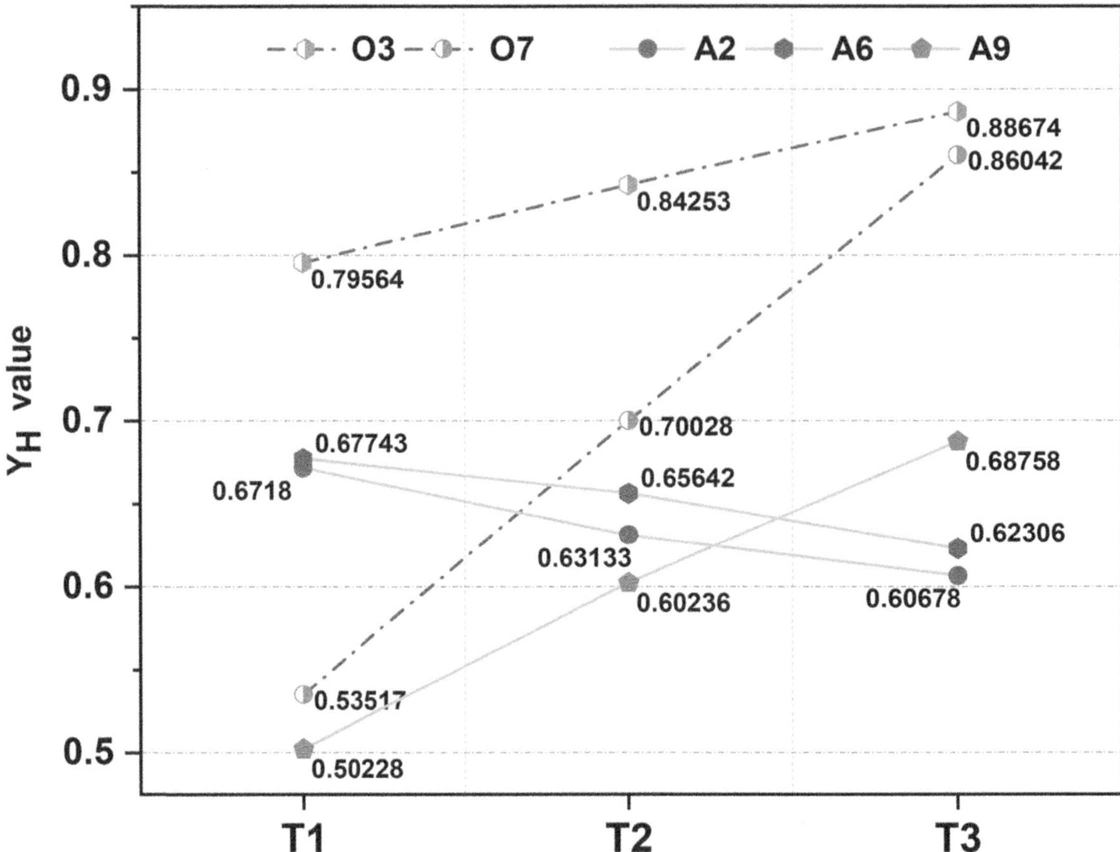

Figure 5 | Y_H values for O3, O7, A2, A6, and A9 at different temperatures.

sample sequence information and diversity index are listed in Supplementary material, Table A4. Figure 6 showed the relative abundance at the phylum and genus levels. A total of 47 different phyla were detected at the three temperatures, with Proteobacteria (21–45%), Chloroflexi (10–32%), Bacteroidota (2–21%), and Actinobacteriota (3–14%) being the top four dominant groups. The sum of these four groups of bacteria in each reactor was above 62%. Chloroflexi is potent in promoting sludge granulation and strengthening of its internal structure. In this study Chloroflexi abundance increased with decreasing temperature in the aerobic reactor and decreased in the anoxic reactor. DO concentrations in the reactor change under low-temperature conditions. It has been shown that when redox substrates (i.e., oxygen and COD) are non-limiting throughout the biofilm, the heterotrophic biofilm quickly grows in a low-density structure (Li *et al.* 2023), whereas nitrifying biofilms tend to already grow at high densities, even at high substrate concentrations (Winkler *et al.* 2013). The occurrence of this phenomenon may be linked to the change in the density of the biofilm with temperature. Additionally, Firmicutes increased in abundance with falling temperature (T2 > T3 > T1), and due to their increased resistance to extreme environments, they ensured the stable operation of the system in low and medium temperature conditions.

Further, the functional genus of aerobic reactors was analyzed and discussed (Figure 6(b)). *Nitrosomonas* was the dominant AOB genus and *Nitrospira*, *Nitrolancea*, and *Candidatus_Nitrotoga* were the dominant NOB genera in the aerobic reactor. *Nitrospira* and *Nitrosomonas* were the main AOB and NOB identified in wastewater treatment plants. *Nitrospira* abundance reached 7% (T1) in the O10-MBBR. The abundance of *Nitrosomonas* with autotrophic ammonia oxidation capacity was significantly higher at T3 than at T1. *Candidatus_Nitrotoga*, a cold-tolerant genus of nitrifying bacteria, showed optimum growth at T3 and was detected in each reactor at T3, and the abundance in both O4-MBBR and O8-MBBR was about 1%, making it the dominant nitrifying genus after *Nitrospira* in the aerobic reactor.

For anoxic reactors, *Denitratisoma*, *Flavobacterium*, *Simplicispira*, *Thiothrix*, and *Hydrogenophaga* were the main functional genera of denitrifying bacteria. Overall, the abundance of denitrifying genera in the system increased significantly

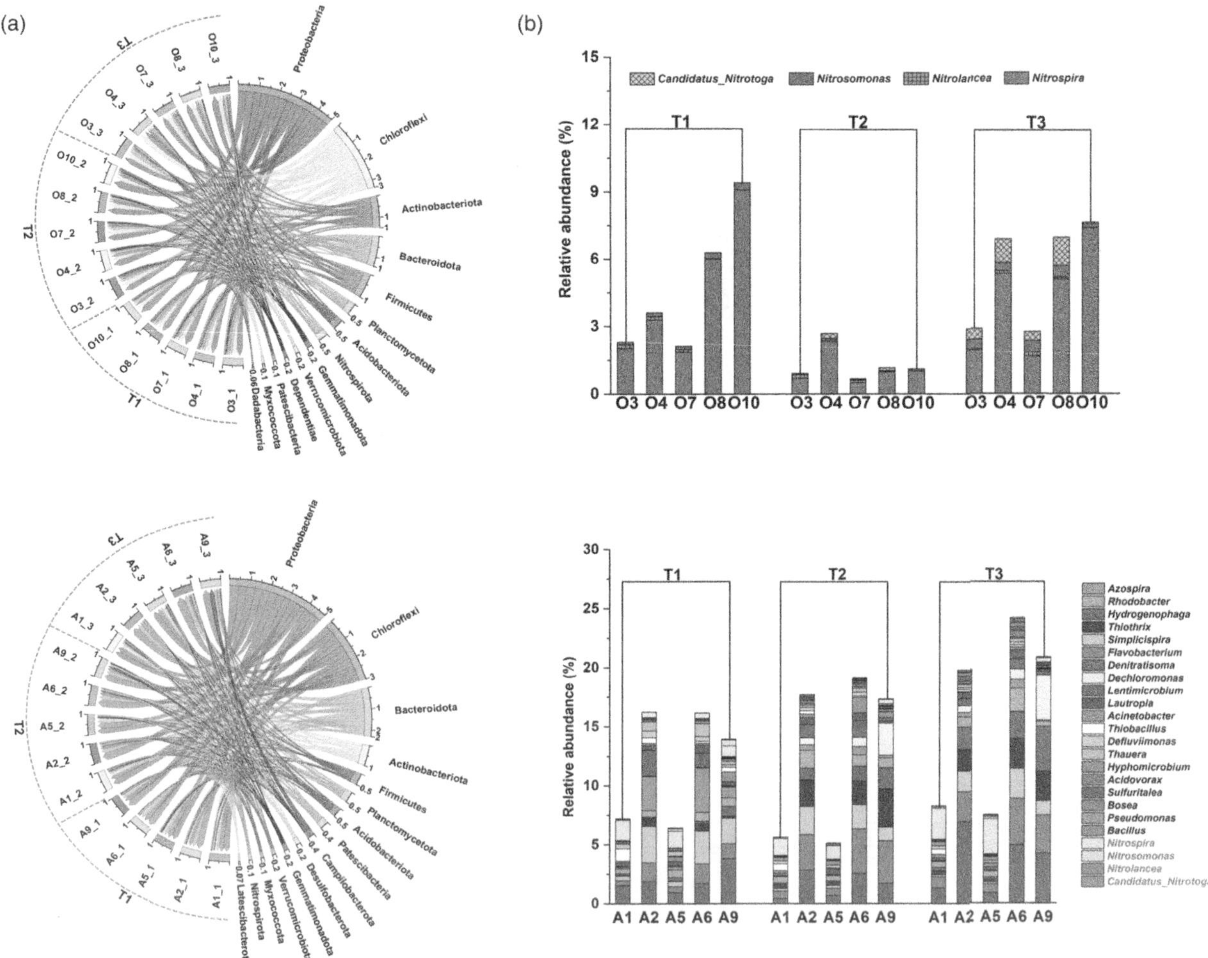

Figure 6 | The microbial community analysis of the multi-stage A/O-MBBR system at different temperatures. (a) Functional microorganisms at the phylum level and (b) functional microorganisms at the genus level.

with decreasing temperature, with increased abundance in the post-anoxic reactors (A2 and A6) compared to that in the pre-anoxic reactors (A1 and A5). These findings are consistent with the X_H data and the respiration rate tests, indicating that they were closely related to each other. Conversely, *Thauera, Rhodobacter, Acinetobacter, Flavobacterium, Thiothrix, Lautropia*, and *Acidovorax* were more abundant in the post-anoxic denitrification reactors, which was different from Zhou *et al.* (2022), who reported that *Rhodobacter, Acinetobacter*, and *Acidovorax* were significantly more abundant in the pre-anoxic reactors. This disparity may be related to the inlet point of the system. *Thauera* and *Flavobacterium* were reported to have the ability of aerobic denitrification and heterotrophic nitrification (Hong *et al.* 2020), with little difference in abundance among reactors. In our study, the abundance of *Azospira* and *Thauera* declined with decreasing temperature. For instance, the relative abundance of *Azospira* and *Thauera* decreased from 4 to 1% and 1 to 0.3% in A6-MBBR, respectively, when the temperature dropped from T1 to T3. *Dechloromonas* has been shown to play an important role in simultaneous nitrification denitrification (SND) (Fan *et al.* 2022), with its levels increasing with decreasing temperature (T3: 0.6% (A2), 0.8% (A6), 4% (A9)), indicating that SND was more likely to occur under low temperature. Furthermore, the *Dechloromonas* were much higher in A9-MBBR than other anoxic reactors, echoing a previous report (Jia *et al.* 2020) indicating that its levels decreased under inadequate carbon sources. Additionally, a higher abundance of *Nitrospira* was observed in the A1-MBBR and A5-MBBR (A1: 1–3%; A5: 1–3%), which corresponded to the more X_A in them. *Nitrospira* was also discovered in A2, A6, and A9 (T1), while being almost absent from A2 and A6 (T2 and T3). This might be explained by the fact that saturated DO was elevated at low temperatures, which was more conducive to the growth of heterotrophic biomass.

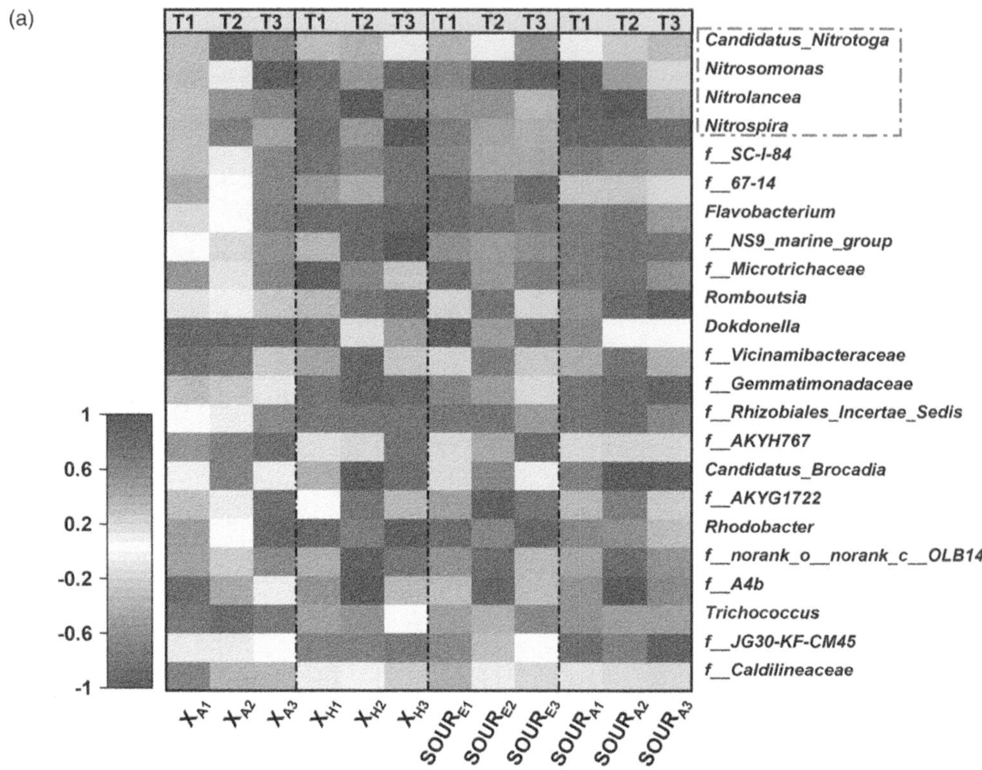

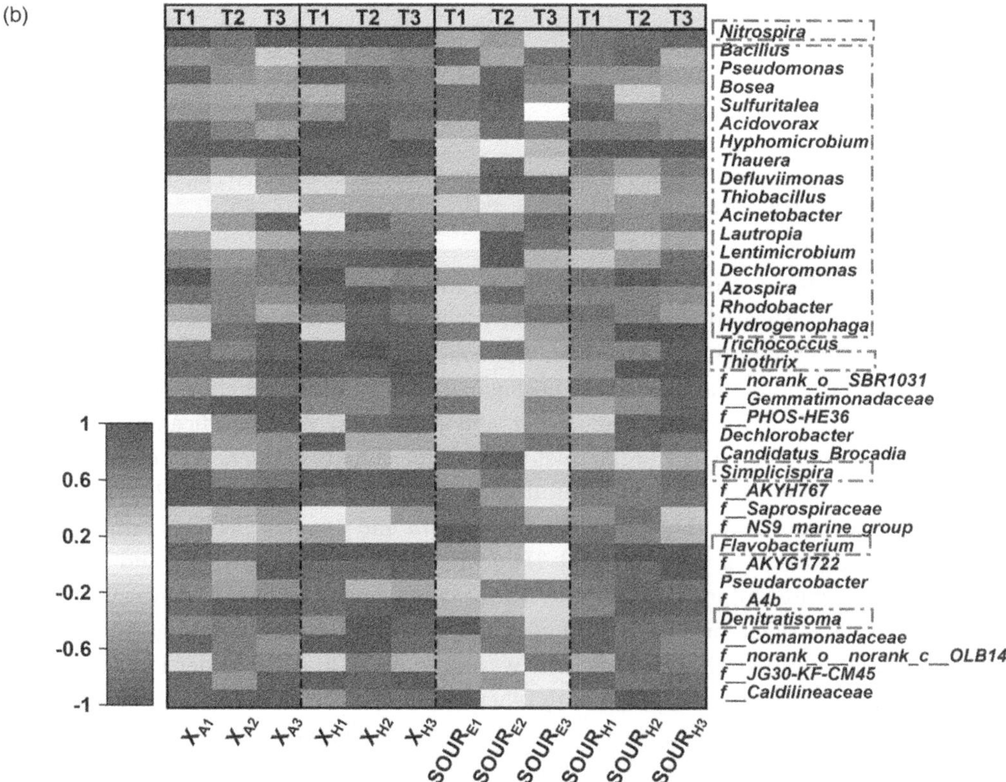

Figure 7 | Correlation analysis of microbial communities with X_A, X_H, and SOUR in aerobic reactors (a) and anoxic reactors (b).

3.5. Correlation analysis of microbial community with X_A, X_H, and SOUR

Our previously presented results indicate a strong link between microbial bacteria, active biomass, and respiration rate. Pearson correlations were calculated for active biomass (X_A, X_H) and SOUR ($SOUR_E$, $SOUR_A$, and $SOUR_H$) at the three temperature conditions with the top 20 terms based on abundance and the functional genus of the anoxic and aerobic reactors (Figure 7).

Nitrospira in the aerobic reactor showed a significant positive correlation with $SOUR_A$ at all temperatures, with a maximum positive correlation coefficient of $0.8216 (p < 0.01)$. The correlation coefficient between *Nitrosomonas* and $SOUR_A$ declined with decreasing temperature $(\rho = 0.8884$ (T1); $\rho = 0.4542$ (T2); $\rho = -0.1982$ (T3)) $(p < 0.05)$. Alternatively, there was a negative correlation between *Nitrolancea* and $SOUR_A$, with a maximum negative correlation coefficient of -0.972; however, the correlation fluctuated considerably due to low temperatures. *Candidatus_Nitrotoga* showed a weaker correlation with $SOUR_A$, probably related to the genus occurring only at low temperatures and being less representative in terms of reactive nitrification performance. Among the dominant genus, *JG30-KF-CM45*, *Rhizobiales_Incertae_Sedis*, and *Romboutsia* had a significant positive correlation with $SOUR_A$ $(p < 0.001)$, with correlation coefficients above 0.5. *Gemmatimonadaceae* had a significant negative correlation with $SOUR_A$ $(\rho = -0.6776-- 0.9893)$ $(p < 0.001)$, and the correlation was further strengthened with decreasing temperature. Except for *Dokdonella* $(p < 0.01)$ and *Trichococcus* $(p < 0.001)$, which showed significant positive correlations with X_A $(\rho > 0.7)$, the correlations between other genera and X_A were weak, while most genera in aerobic and anoxic reactors showed significant correlations with X_H, which may have a certain relationship with the base of the active biomass type. Additionally, the correlation between the genus and X_H as well as $SOUR_E$ showed a strong consistency in the aerobic reactor, indicating that the biomass decay in the system was mainly dominated by heterotrophic biomass.

Except *Hyphomicrobium* and *Thiobacillus*, all other denitrifying genera showed a positive relationship with $SOUR_H$ and X_H, and a negative relationship with X_A in anoxic reactors. *Thiothrix* and *Hydrogenophaga* showed enhanced positive correlations with $SOUR_H$ with decreasing temperature $(\rho = 0.51$ and 0.63 (T1); $\rho = 0.99$ and 1.00 (T2); $\rho = 0.98$ and 0.97 (T3), respectively) $(p < 0.05)$. The SND-related *Dechloromonas* showed a positive correlation with X_H $(p < 0.01)$ and $SOUR_H$ $(p < 0.05)$ and a negative correlation with X_A $(p < 0.05)$ (T1). Its correlation with active biomass weakened with decreasing temperature, while its correlation with the respiration rate was slightly increased. *Flavobacterium* (aerobic denitrification capacity) showed a significant positive correlation with X_H $(\rho > 0.97)$ $(p < 0.01)$ and $SOUR_H$ $(p < 0.05)$. The correlation between *Thauera* and X_H was more significant $(\rho = 0.83–0.86)$ $(p < 0.05)$. As a typical genus of endogenous denitrifying bacteria using an internal carbon source (Salehi *et al.* 2019), *Dechloromonas* showed a negative correlation with $SOUR_E$ $(\rho = 0.48–0.67)$, while *Defluviicoccus* showed a significant positive correlation with $SOUR_E$ $(\rho = 0.48–0.98)$ $(p < 0.001)$, with more significant correlations at low temperatures. Finally, the most dominant genera in the anoxic reactor showed significant correlations with X_A, X_H, and $SOUR_H$, and were less influenced by overall temperature.

In summary, the correlation between the observed microbial community, the active biomass, and the specific respiration rate reflects the reliability of the respirometric method for characterizing the microbial performance of the multi-stage A/O-MBBR. As the first study to propose a correlation between microbial community, active biomass, and respiration rate, the method is yet to be validated by similar studies.

4. CONCLUSIONS

In this study, the active biomass distribution, Y_H of a multi-stage MBBR with nitrogen removal were measured in different conditions. The correlation between SOUR and nitrification as well as denitrification activity was also investigated. The following conclusions were drawn:

- The total active biomass in the multi-stage A/O-MBBR system was 0.71–1.68 g COD/m^2 for the aerobic reactor and 0.39–1.44 g COD/m^2 for the anoxic reactor.
- The increased active biomass at low temperatures can to some extent compensate for the reduced biofilm activity per unit mass.
- The Y_H values for the aerobic reactors were 0.82–0.89 at 10–13 °C, which were significantly higher than the recommended value of the activated sludge model ($Y_H = 0.67$). However, the anoxic reactors had Y_H value of 0.61–0.69, which were comparable to $Y_H = 0.67$.

• The correlation coefficient between *Nitrospira* and $SOUR_A$ was 0.82. *Thiothrix, Hydrogenophaga, Dechloromonas, Flavobacterium* and other denitrifying genera showed significant correlation with $SOUR_H$ and X_H.

The results of the correlation analysis between the community and SOUR also concluded that the respiration tests applied in this study can function as a simple-to-apply and reliable method for nitrification and denitrification trouble shooting at WWTPs.

FUNDING

This work was funded by the National Natural Science Foundation of China (grant number: 51908303) and the Research Council of Norway (grant number: 310074/G10).

AUTHOR CONTRIBUTIONS

N.C. investigated the study, did data curation, wrote the original draft, did formal analysis and prepared the methodology. X.W. conceptualized the study, investigated and prepared the methodology, wrote the draft and acquired funds. M.H. investigated the study and did data curation. Z.M. wrote the original draft, revised, and editing the article. H.R. validated the study. X.B. investigated and supervised the study.

DATA AVAILABILITY STATEMENT

All relevant data are included in the paper or its Supplementary Information.

CONFLICT OF INTEREST

The authors declare there is no conflict.

REFERENCES

Almomani, F. A., Delatolla, R. & Örmeci, B. 2014 Field study of moving bed biofilm reactor technology for post-treatment of wastewater lagoon effluent at 1°C. *Environ. Technol.* **35**, 1596–1604. https://doi.org/10.1080/09593330.2013.874500.

Amin, M., Farokhzaddeh, H., Fatehizadeh, A., Ghasemian, M., Moradpour, H., Nikaeen, M., Shafiea, A., Molayi, R. & Sabouri, A. 2013 Biodegradation performance of anaerobic sequencing batch biofilm reactor for oil with polychlorinated biphenyls. *Int. J. Environ. Health Eng.* **2**, 19. https://doi.org/10.4103/2277-9183.110175.

Antoniou, P., Hamilton, J., Koopman, B., Jain, R., Holloway, B., Lyberatos, G. & Svoronos, S. A. 1990 Effect of temperature and pH on the effective maximum specific growth rate of nitrifying bacteria. *Water Res.* **24**, 97–101. https://doi.org/10.1016/0043-1354(90)90070-M.

Ashkanani, A., Almomani, F., Khraisheh, M., Bhosale, R., Tawalbeh, M. & AlJaml, K. 2019 Bio-carrier and operating temperature effect on ammonia removal from secondary wastewater effluents using moving bed biofilm reactor (MBBR). *Sci. Total Environ.* **693**, 133425. https://doi.org/10.1016/j.scitotenv.2019.07.231.

Avcioglu, E., Orhon, D. & Sözen, S. 1998 A new method for the assessment of heterotrophic endogenous respiration rate under aerobic and anoxic conditions. *Water Sci. Technol.* **38**, 95–103. https://doi.org/10.2166/wst.1998.0795.

Bina, B., Mohammadi, F., Amin, M. M., Pourzamani, H. R. & Yavari, Z. 2018 Evaluation of the effects of alkylPhenolic compounds on kinetic coefficients and biomass activity in MBBR by means of respirometric techniques. *Chinese J. Chem. Eng.* **26**, 822–829. https://doi.org/10.1016/j.cjche.2017.07.024.

Collivignarelli, M. C., Bertanza, G., Abbà, A., Torretta, V. & Katsoyiannis, I. A. 2019 Wastewater treatment by means of thermophilic aerobic membrane reactors: respirometric tests and numerical models for the determination of stoichiometric/kinetic parameters. *Environ. Technol.* **40**, 182–191. https://doi.org/10.1080/09593330.2017.1384070.

Delatolla, R., Tufenkji, N., Comeau, Y., Gadbois, A., Lamarre, D. & Berk, D. 2010 Investigation of laboratory-scale and pilot-scale attached growth ammonia removal kinetics at cold temperature and low influent carbon. *Water Qual. Res. J.* **45**, 427–436. https://doi.org/10.2166/wqrj.2010.042.

di Biase, A., Kowalski, M. S., Devlin, T. R. & Oleszkiewicz, J. A. 2021 Modeling of the attached and suspended biomass fractions in a moving bed biofilm reactor. *Chemosphere* **275**, 129937. https://doi.org/10.1016/j.chemosphere.2021.129937.

Di Trapani, D., Capodici, M., Cosenza, A., Di Bella, G., Mannina, G., Torregrossa, M. & Viviani, G. 2011 Evaluation of biomass activity and wastewater characterization in a UCT-MBR pilot plant by means of respirometric techniques. *Desalination* **269**, 190–197. https://doi.org/10.1016/j.desal.2010.10.061.

Du, J., Qv, W., Niu, Y., Qv, M., Jin, K., Xie, J. & Li, Z. 2022 Nanoplastic pollution inhibits stream leaf decomposition through modulating microbial metabolic activity and fungal community structure. *J. Hazard. Mater.* **424**, 127392. https://doi.org/10.1016/j.jhazmat.2021.127392.

Fan, Y., Zhang, M., Cheng, J., Yong, D., Ji, J., Wu, Q. & He, C. 2022 Elucidating nitrifying performance, nitrite accumulation and microbial community in a three-stage plug flow moving bed biofilm reactor (PF – MBBR). *Chemosphere* **297**, 134087. https://doi.org/10.1016/j.chemosphere.2022.134087.

Fathali, D., Rashidi Mehrabadi, A., Mirabi, M. & Alimohammadi, M. 2019 Investigation on nitrogen removal performance of an enhanced post-anoxic membrane bioreactor using disintegrated sludge as a carbon source: an experimental study. *J. Environ. Chem. Eng.* **7**, 103445. https://doi.org/10.1016/j.jece.2019.103445.

Fernandes, H., Hoffmann, H., Antonio, R. V. & Costa, R. H. R. 2014 The role of microorganisms in a full-scale sequencing batch reactor under low aeration and different cycle times. *Water Environ. Res.* **86**, 800–809. https://doi.org/10.2175/106143013X13807328848450.

Ferrai, M., Guglielmi, G. & Andreottola, G. 2010 Modelling respirometric tests for the assessment of kinetic and stoichiometric parameters on MBBR biofilm for municipal wastewater treatment. *Environ. Model. Softw.* **25**, 626–632. https://doi.org/10.1016/j.envsoft.2009.05.005.

Fu, X., Hou, R., Yang, P., Qian, S., Feng, Z., Chen, Z., Wang, F., Yuan, R., Chen, H. & Zhou, B. 2022 Application of external carbon source in heterotrophic denitrification of domestic sewage: a review. *Sci. Total Environ.* **817**, 153061. https://doi.org/10.1016/j.scitotenv.2022.153061.

Ghafari, S., Hasan, M. & Aroua, M. K. 2008 Bio-electrochemical removal of nitrate from water and wastewater – a review. *Bioresour. Technol.* **99**, 3965–3974. https://doi.org/10.1016/j.biortech.2007.05.026.

Gu, Q., Sun, T., Wu, G., Li, M. & Qiu, W. 2014 Influence of carrier filling ratio on the performance of moving bed biofilm reactor in treating coking wastewater. *Bioresour. Technol.* **166**, 72–78. https://doi.org/10.1016/j.biortech.2014.05.026.

Gujer, W. 2006 Activated sludge modelling: past, present and future. *Water Sci. Technol.* **53**, 111–119. https://doi.org/10.2166/wst.2006.082.

Hayet, C., Saida, B.-A., Youssef, T. & Hédi, S. 2016 Study of biodegradability for municipal and industrial Tunisian wastewater by respirometric technique and batch reactor test. *Sustain. Environ. Res.* **26**, 55–62. https://doi.org/10.1016/j.serj.2015.11.001.

He, S., Wang, Y., Li, C., Li, Y. & Zhou, J. 2018 The nitrogen removal performance and microbial communities in a two-stage deep sequencing constructed wetland for advanced treatment of secondary effluent. *Bioresour. Technol.* **248**, 82–88. https://doi.org/10.1016/j.biortech.2017.06.150.

Helbling, D. E., Johnson, D. R., Lee, T. K., Scheidegger, A. & Fenner, K. 2015 A framework for establishing predictive relationships between specific bacterial 16S rRNA sequence abundances and biotransformation rates. *Water Res.* **70**, 471–484. https://doi.org/10.1016/j.watres.2014.12.013.

Henze, M. Gujer, W. Mino, T. & van Loosedrecht, M. 2000 *Activated Sludge Models ASM1, ASM2, ASM2d and ASM3*. IWA Scientific and Technical Report No. 9. IWA Publishing, London, UK.

Hong, P., Wu, X., Shu, Y., Wang, C., Tian, C., Wu, H. & Xiao, B. 2020 Bioaugmentation treatment of nitrogen-rich wastewater with a denitrifier with biofilm-formation and nitrogen-removal capacities in a sequencing batch biofilm reactor. *Bioresour. Technol.* **303**, 122905. https://doi.org/10.1016/j.biortech.2020.122905.

Hurse, T. J. & Connor, M. A. 1999 Nitrogen removal from wastewater treatment lagoons. *Water Sci. Technol.* **39**, 191–198. https://doi.org/10.2166/wst.1999.0296.

Jia, Y., Zhou, M., Chen, Y., Hu, Y. & Luo, J. 2020 Insight into short-cut of simultaneous nitrification and denitrification process in moving bed biofilm reactor: effects of carbon to nitrogen ratio. *Chem. Eng. J.* **400**, 125905. https://doi.org/10.1016/j.cej.2020.125905.

Johnson, D. R., Helbling, D. E., Lee, T. K., Park, J., Fenner, K., Kohler, H.-P. E. & Ackermann, M. 2015 Association of biodiversity with the rates of micropollutant biotransformations among full-scale wastewater treatment plant communities. *Appl. Environ. Microbiol.* **81**, 666–675. https://doi.org/10.1128/AEM.03286-14.

Joss, A., Zabczynski, S., Göbel, A., Hoffmann, B., Löffler, D., McArdell, C. S., Ternes, T. A., Thomsen, A. & Siegrist, H. 2006 Biological degradation of pharmaceuticals in municipal wastewater treatment: proposing a classification scheme. *Water Res.* **40**, 1686–1696. https://doi.org/10.1016/j.watres.2006.02.014.

Karanasios, K. A., Vasiliadou, I. A., Pavlou, S. & Vayenas, D. V. 2010 Hydrogenotrophic denitrification of potable water: a review. *J. Hazard. Mater.* **180**, 20–37. https://doi.org/10.1016/j.jhazmat.2010.04.090.

Kumar, M., Lee, P.-Y., Fukusihma, T., Whang, L.-M. & Lin, J.-G. 2012 Effect of supplementary carbon addition in the treatment of low C/N high-technology industrial wastewater by MBR. *Bioresour. Technol.* **113**, 148–153. https://doi.org/10.1016/j.biortech.2011.12.102.

Leyva-Díaz, J. C., Martín-Pascual, J., González-López, J., Hontoria, E. & Poyatos, J. M. 2013 Effects of scale-up on a hybrid moving bed biofilm reactor – membrane bioreactor for treating urban wastewater. *Chem. Eng. Sci.* **104**, 808–816. https://doi.org/10.1016/j.ces.2013.10.004.

Li, C., Gu, Z., Zhu, S. & Liu, D. 2020 17β-Estradiol removal routes by moving bed biofilm reactors (MBBRs) under various C/N ratios. *Sci. Total Environ.* **741**, 140381. https://doi.org/10.1016/j.scitotenv.2020.140381.

Li, J., Zheng, L., Ye, C., Ni, B., Wang, X. & Liu, H. 2021 Evaluation of an intermittent-aeration constructed wetland for removing residual organics and nutrients from secondary effluent: performance and microbial analysis. *Bioresour. Technol.* **329**, 124897. https://doi.org/10.1016/j.biortech.2021.124897.

Li, M., Perez-Calleja, P., Kim, B., Picioreanu, C. & Nerenberg, R. 2023 Unique stratification of biofilm density in heterotrophic membrane-aerated biofilms: an experimental and modeling study. *Chemosphere* **327**, 138501. https://doi.org/10.1016/j.chemosphere.2023.138501.

Luan, Y.-N., Yin, Y., An, Y., Zhang, F., Wang, X., Zhao, F., Xiao, Y. & Liu, C. 2022 Investigation of an intermittently-aerated moving bed biofilm reactor in rural wastewater treatment under low dissolved oxygen and C/N condition. *Bioresour. Technol.* **358**, 127405. https://doi.org/10.1016/j.biortech.2022.127405.

Ma, Q., Qu, Y., Shen, W., Zhang, Z., Wang, J., Liu, Z., Li, D., Li, H. & Zhou, J. 2015 Bacterial community compositions of coking wastewater treatment plants in steel industry revealed by illumina high-throughput sequencing. *Bioresour. Technol.* **179**, 436–443. https://doi.org/10.1016/j.biortech.2014.12.041.

Mansouri, A. M. 2014 Kinetic evaluation of simultaneous CNP removal in an up-flow aerobic/Anoxic sludge fixed film (UAASFF) bioreactor. *Iran. J. Energy Environ.* **5**. https://doi.org/10.5829/idosi.ijee.2014.05.03.12.

Morgan-Sagastume, F. 2018 Biofilm development, activity and the modification of carrier material surface properties in moving-bed biofilm reactors (MBBRs) for wastewater treatment. *Crit. Rev. Environ. Sci. Technol.* **48**, 439–470. https://doi.org/10.1080/10643389.2018.1465759.

Ochoa, J. C., Colprim, J., Palacios, B., Paul, E. & Chatellier, P. 2002 Active heterotrophic and autotrophic biomass distribution between fixed and suspended systems in a hybrid biological reactor. *Water Sci. Technol.* **46**, 397–404. https://doi.org/10.2166/wst.2002.0507.

Ooi, G. T. H., Tang, K., Chhetri, R. K., Kaarsholm, K. M. S., Sundmark, K., Kragelund, C., Litty, K., Christensen, A., Lindholst, S., Sund, C., Christensson, M., Bester, K. & Andersen, H. R. 2018 Biological removal of pharmaceuticals from hospital wastewater in a pilot-scale staged moving bed biofilm reactor (MBBR) utilising nitrifying and denitrifying processes. *Bioresour. Technol.* **267**, 677–687. https://doi.org/10.1016/j.biortech.2018.07.077.

Pelaz, L., Gómez, A., Letona, A., Garralón, G. & Fdz-Polanco, M. 2018 Nitrogen removal in domestic wastewater. effect of nitrate recycling and COD/N ratio. *Chemosphere* **212**, 8–14. https://doi.org/10.1016/j.chemosphere.2018.08.052.

Piculell, M., Welander, T. & Jönsson, K. 2014 Organic removal activity in biofilm and suspended biomass fractions of MBBR systems. *Water Sci. Technol.* **69**, 55–61. https://doi.org/10.2166/wst.2013.552.

Polesel, F., Torresi, E., Loreggian, L., Casas, M. E., Christensson, M., Bester, K. & Plósz, B. G. 2017 Removal of pharmaceuticals in pre-denitrifying MBBR – influence of organic substrate availability in single- and three-stage configurations. *Water Res.* **123**, 408–419. https://doi.org/10.1016/j.watres.2017.06.068.

Revilla, M., Galán, B. & Viguri, J. R. 2016 An integrated mathematical model for chemical oxygen demand (COD) removal in moving bed biofilm reactors (MBBR) including predation and hydrolysis. *Water Res.* **98**, 84–97. https://doi.org/10.1016/j.watres.2016.04.003.

Salehi, S., Cheng, K. Y., Heitz, A. & Ginige, M. P. 2019 Simultaneous nitrification, denitrification and phosphorus recovery (SNDPr) – an opportunity to facilitate full-scale recovery of phosphorus from municipal wastewater. *J. Environ. Manage.* **238**, 41–48. https://doi.org/10.1016/j.jenvman.2019.02.063.

Sánchez-Zurano, A., Rossi, S., Fernández-Sevilla, J. M., Acién-Fernández, G., Molina-Grima, E. & Ficara, E. 2022 Respirometric assessment of bacterial kinetics in algae-bacteria and activated sludge processes. *Bioresour. Technol.* **352**, 127116. https://doi.org/10.1016/j.biortech.2022.127116.

Strotmann, U. J., Geldem, A., Kuhn, A., Gendig, C. & Klein, S. 1999 Evaluation of a respirometric test method to determine the heterotrophic yield coefficient of activated sludge bacteria. *Chemosphere* **38**, 3555–3570. https://doi.org/10.1016/S0045-6535(98)00569-4.

Sun, F., Lv, X., Li, J., Peng, Z., Li, P. & Shao, M. 2014 Activated sludge filterability improvement by nitrifying bacteria abundance regulation in an adsorption membrane bioreactor (Ad-MBR). *Bioresour. Technol.* **170**, 230–238. https://doi.org/10.1016/j.biortech.2014.07.092.

Torresi, E., Escolà Casas, M., Polesel, F., Plósz, B. G., Christensson, M. & Bester, K. 2017 Impact of external carbon dose on the removal of micropollutants using methanol and ethanol in post-denitrifying moving bed biofilm reactors. *Water Res.* **108**, 95–105. https://doi.org/10.1016/j.watres.2016.10.068.

Torresi, E., Gülay, A., Polesel, F., Jensen, M. M., Christensson, M., Smets, B. F. & Plósz, B. G. 2018 Reactor staging influences microbial community composition and diversity of denitrifying MBBRs- implications on pharmaceutical removal. *Water Res.* **138**, 333–345. https://doi.org/10.1016/j.watres.2018.03.014.

Vanrolleghem, P. A., Spanjers, H., Petersen, B., Ginestet, P. & Takacs, I. 1999 Estimating (combinations of) activated sludge model no. 1 parameters and components by respirometry. *Water Sci. Technol.* **39**, 195–214. https://doi.org/10.2166/wst.1999.0042.

Vasiliadou, I. A., Molina, R., Martinez, F., Melero, J. A., Stathopoulou, P. M. & Tsiamis, G. 2018 Toxicity assessment of pharmaceutical compounds on mixed culture from activated sludge using respirometric technique: the role of microbial community structure. *Sci. Total Environ.* **630**, 809–819. https://doi.org/10.1016/j.scitotenv.2018.02.095.

Wang, J. & Chu, L. 2016 Biological nitrate removal from water and wastewater by solid-phase denitrification process. *Biotechnol. Adv.* **34**, 1103–1112. https://doi.org/10.1016/j.biotechadv.2016.07.001.

Wang, X., Bi, X., Hem, L. J. & Ratnaweera, H. 2018 Microbial community composition of a multi-stage moving bed biofilm reactor and its interaction with kinetic model parameters estimation. *J. Environ. Manage.* **218**, 340–347. https://doi.org/10.1016/j.jenvman.2018.04.015.

Wang, X., Zhao, J., Yu, D., Du, S., Yuan, M. & Zhen, J. 2019 Evaluating the potential for sustaining mainstream anammox by endogenous partial denitrification and phosphorus removal for energy-efficient wastewater treatment. *Bioresour. Technol.* **284**, 302–314. https://doi.org/10.1016/j.biortech.2019.03.127.

Winkler, M. K., Kleerebezem, R., Strous, M., Chandran, K. & van Loosdrecht, M. C. M. 2013 Factors influencing the density of aerobic granular sludge. *Appl. Microbiol. Biotechnol.* **97** (16), 7459–7468. https://doi.org/10.1007/s00253-012-4459-4.

Wu, S., Kuschk, P., Brix, H., Vymazal, J. & Dong, R. 2014 Development of constructed wetlands in performance intensifications for wastewater treatment: a nitrogen and organic matter targeted review. *Water Res.* **57**, 40–55. https://doi.org/10.1016/j.watres.2014.03.020.

Young, B., Delatolla, R., Ren, B., Kennedy, K., Laflamme, E. & Stintzi, A. 2016 Pilot-scale tertiary MBBR nitrification at 1 °C: characterization of ammonia removal rate, solids settleability and biofilm characteristics. *Environ. Technol.* **37**, 2124–2132. https://doi.org/10.1080/09593330.2016.1143037.

Young, B., Delatolla, R., Kennedy, K., Laflamme, E. & Stintzi, A. 2017a Low temperature MBBR nitrification: microbiome analysis. *Water Res.* **111**, 224–233. https://doi.org/10.1016/j.watres.2016.12.050.

Young, B., Delatolla, R., Kennedy, K., LaFlamme, E. & Stintzi, A. 2017b Post carbon removal nitrifying MBBR operation at high loading and exposure to starvation conditions. *Bioresour. Technol.* **239**, 318–325. https://doi.org/10.1016/j.biortech.2017.05.024.

Zhang, M., Yu, M., Wang, Y., He, C., Pang, J. & Wu, J. 2019 Operational optimization of a three-stage nitrification moving bed biofilm reactor (NMBBR) by obtaining enriched nitrifying bacteria: nitrifying performance, microbial community, and kinetic parameters. *Sci. Total Environ.* **697**, 134101. https://doi.org/10.1016/j.scitotenv.2019.134101.

Zhang, H., Gao, Z., Shi, M. & Fang, S. 2020 Soil bacterial diversity and its relationship with soil CO_2 and mineral composition: a case study of the Laiwu experimental site. *Int. J. Environ. Res. Public Health* **17**, 5699. https://doi.org/10.3390/ijerph17165699.

Zhou, X., Bi, X., Fan, X., Yang, T., Wang, X., Chen, S., Cheng, L., Zhang, Y., Zhao, W., Zhao, F., Nie, S. & Deng, X. 2022 Performance and bacterial community analysis of a two-stage A/O-MBBR system with multiple reactors for biological nitrogen removal. *Chemosphere* **303**, 135195. https://doi.org/10.1016/j.chemosphere.2.135195.

First received 20 January 2023; accepted in revised form 3 May 2023. Available online 22 May 2023

© 2023 The Authors

doi: 10.2166/wrd.2023.062

Degradation of sulfonamide antibiotics in the rhizosphere of two dominant plants in Huixian karst wetland, Guangxi, China

Kun Dong [a], WuBin Wang[a], Min Li[b], Xinyu Zhou[b], Yutong Huang[b], Guozhi Zhou[b], Yufeng Xu[b], Dunqiu Wang[b] and Hai-Xiang Li[b],*

[a] Collaborative Innovation Center for Water Pollution Control and Water Safety in Karst Area, Guilin University of Technology, Guilin 541004, China
[b] Guangxi Key Laboratory of Environmental Pollution Control Theory and Technology, Guilin University of Technology, Guilin 541004, China
*Corresponding author. E-mail: lihaixiang0627@163.com

KD, 0000-0001-7006-274X

ABSTRACT

In this work, *Phragmites australis* and *Vallisneria natans* were selected as the research objects and were cultured for 10 d under 0.10 µg L^{-1} sulfadiazine (SD) stress in a simulated surface flow wetland reactor. SD degradation was conducted at pH = 7 and 25 °C for 96 h. Each plant group conformed to the first-order kinetic model of degradation, and the degradation rate increased with time, reaching the maximum at 96 h. At 96 h, the degradation rate of *P. australis* communities was higher than that of *V. natans*. SD metabolites showed that the degradation pathways in the plant rhizosphere were mainly hydroxylation, aminolation, and S–N bond cleavage. In the analysis of rhizosphere bacterial community structure, the bacterial phyla that could degrade antibiotics accounted for a large proportion. Compared with before degradation, the dominant phylum and genus did not change after degradation (96 h), but their abundance changed to varying degrees, and new genera appeared in the *P. australis* group. This research provides a reference for the degradation of antibiotics in karst areas and new information on the mechanism of SA degradation in the plant rhizosphere.

Key words: bacterial community, degradation, karst wetland, rhizosphere secretion, sulfonamides antibiotics

HIGHLIGHTS

- Under sulfonamide (SD) stress, the contents of organic acid esters in rhizosphere exudates of *P. australis* and *V. natans* increased.
- New genera were produced in rhizosphere soil of *P. australis* group during SD degradation.
- Antibiotics in karst wetlands may exist in the form of complexes and be adsorbed in soil or sediments.
- After SD degradation, the diversity of bacteria decreased, but the total number of bacteria increased.

1. INTRODUCTION

Sulfonamide antibiotics (SAs) are synthetic antibacterial agents derived from sulfanilic acid. They are often added to veterinary feed or drugs in the form of additives to promote animal growth and disease prevention (Man *et al.* 2019). SAs are only partially metabolized in organisms, and 50–90% of the SAs in feces or urine will be discharged into the environment with their original structures intact, thus accumulating in various environmental media, such as surface water, groundwater, and soil (Zhang *et al.* 2014; Cui *et al.* 2020). Using plants to purify SA-polluted water has become a feasible technology (He *et al.* 2016). Phytoremediation is affected by rhizosphere microorganisms, rhizosphere soil contains a large number of SA-degrading bacteria that can use SAs as a designated carbon source for metabolic activities (Chen *et al.* 2016). Compared with oxidation processes, adsorption on activated carbon, and membrane filtration, phytoremediation has the advantages of high efficiency and low cost (Hu *et al.* 2019).

Karst aquifers are important sources of drinking water in many areas of the world. Because of the high hydraulic conductivity and short residence time of karst systems, antibiotics can be transported in the karst pipeline (Hillebrand *et al.* 2015), as a result, karst ecosystems are easily polluted by antibiotics. At present, antibiotics have been detected in the groundwater of karst areas in countries including Switzerland, USA, Germany, and France (Dodgen *et al.* 2017). The degradation of antibiotics in the environment is affected by environmental factors including temperature, pH, and ionic strength (Menció &

Mas-Pla 2019). The hydrochemical characteristics of karst areas are quite different from those of non-karst areas. The long-term corrosion of carbonate rocks easily forms the special environment of 'high calcium (Ca^{2+}), alkali, high dissolved inorganic carbon (HCO_3^-)' found in karst areas (Li *et al.* 2015; Zhang *et al.* 2018). Therefore, the degradation of antibiotics in karst areas may be different from that in non-karst areas. With frequent human production activities (domestic sewage, medical wastewater, aquaculture and livestock farming, etc.), antibiotics, as a trace pollutant in the environment, are continuously discharged into natural water bodies such as wetlands, altering the dynamic balance of antibiotics in the original environment. A study by Conkle *et al.* (2008) showed that at initial concentrations of 4.090 and 0.068 ug L^{-1} for sulfamethoxazole and sulfadiazine, respectively, the non-karst forest wetlands were less effective in removing the two sulfonamides and less effective in treating and diluting the antibiotics.

As the largest karst wetland in China, Huixian wetland plays important ecological roles in water conservation, climate regulation, water purification, biodiversity conservation, and flood storage in Li River. As part of its involvement in urban water recycling, the Huixian wetland is to some extent able to convert engineered reclaimed water into ecologically reclaimed water, a process that is particularly important in controlling the conversion of new pollutants such as antibiotics. It was reported that SAS pollution was detected in surface water, shallow groundwater, water near farms, and pond water of Huixian wetland (Qin *et al.* 2019). Therefore, the management of SAs in Huixian wetland has gradually attracted the attention of Chinese and international scholars. There are many types of vegetation in Huixian wetland. The plant coverage can reach 80–95%, and the dominant plants are *Cladium chinense Nees*, *Typha orientalis Presl*, *Vallisneria natans*, *Phragmites australis*, and *Canna indica* (Tu *et al.* 2019). *P. australis* is an emergent plant with strong resistance, fast growth, high yield, and strong adsorption, and is the preferred plant for purifying water quality (Lambertini *et al.* 2020; Wang *et al.* 2021a). *V. natans* is a perennial submerged plant that can effectively remove nitrogen, phosphorus, and heavy metals in polluted water (Yan *et al.* 2011). The investigation of dominant plant species can be used to assess biodiversity changes and ecosystem functions in wetland systems. It has been reported that the removal rate of SAs in plant systems was significantly higher than that in non-plant systems, and it was found that the degradation of SAs by plants was related to plant accumulation and stress resistance ability, and the degradation of antibiotics in plants was mainly related to rhizosphere microorganisms (Chen *et al.* 2016, 2021). It has also been reported that root exudates can play a very significant role in promoting microbial degradation of antibiotics (Zhi *et al.* 2019), and different kinds of substances in root secretions will increase rhizosphere microbial biomass and activity, and then affect how effectively wetland systems purify pollutants (Du *et al.* 2020).

However, few studies have revealed the degradation efficiency and mechanism of SAs in the rhizosphere of wetland plants in karst areas, and little attention has been paid to changes in the rhizosphere microbial community structure before and after antibiotic degradation. Therefore, this work selected Huixian wetland dominant plants (*P. australis* and *V. natans*) as the research object. In the simulated surface flow wetland reactor, the root exudates were extracted after 10 d of incubation under sulfadiazine (SD) stress. The root exudates were mixed with the rhizosphere soil to explore the degradation mechanism of SD by root exudates, and the composition and changes of bacterial communities in the rhizosphere soil before and after degradation were analyzed by high-throughput sequencing technology to reveal the mechanism of SD degradation in the rhizosphere so as to provide a theoretical basis for antibiotic treatment in Huixian wetland.

2. METHODS

2.1. Sample collection and pretreatment

The two dominant wetland plants (*P. australis* and *V. natans*) used in the experiment were all obtained from the Qixing Wharf Basin of Huixian wetland (25°01′30″–25°11′15″N, 110°08′15″–110°18′00″E) (Figure 1). The sampling time was March 2020, and plants with high density and vigorous growth were selected in an effort to ensure the selection of plants with similar growth characteristics, such as the plant height, number of roots, and leaves. After the collection of plant samples, to ensure that the root system was not damaged, tap water was first used to gently rinse the roots, and the soil, residual leaves, snail shells, and gravel were rinsed away. After washing, the roots and stems were soaked in ultrapure water for 5 min to reduce other pollutants. After soaking, the roots and stems were washed with ultrapure water more than three times. After air-drying, the rhizosphere soil is ground down to 10 and 60 mesh, one is used for degradation experiment (10 mesh), and the other is used to detect the initial physical and chemical properties (60 mesh).

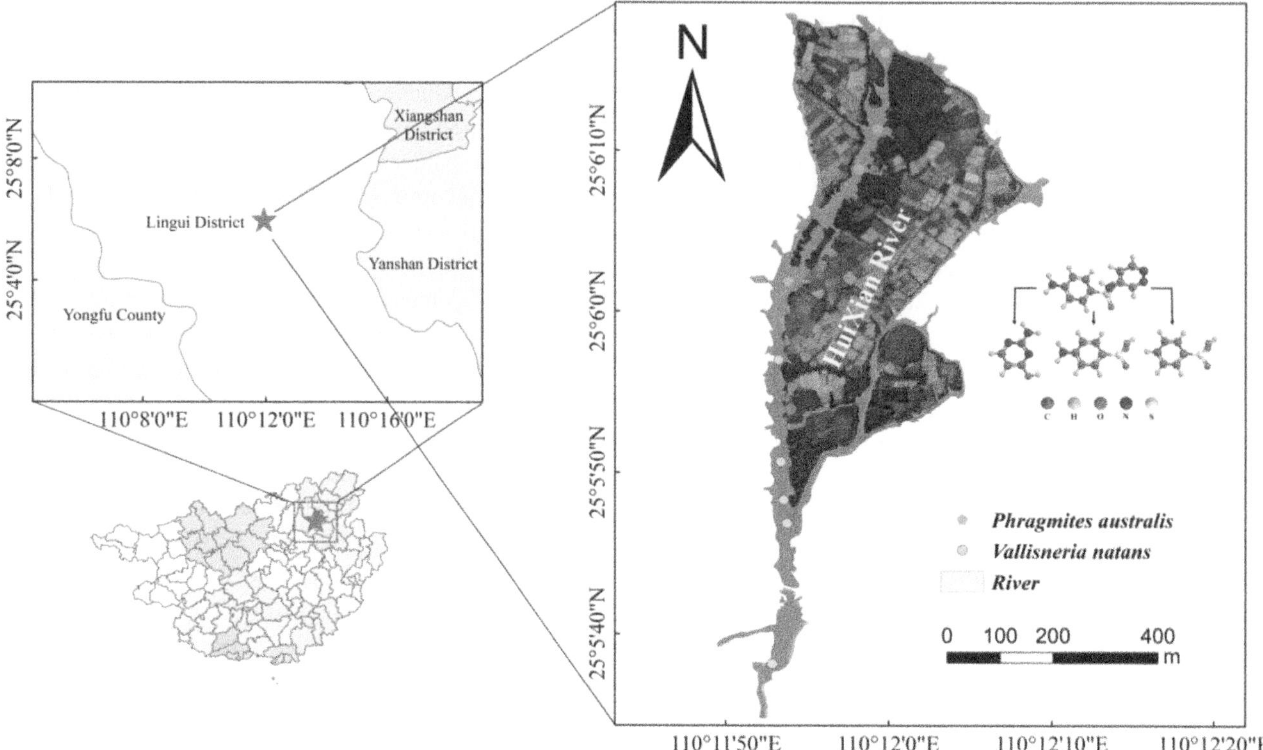

Figure 1 | Information of sampling sites and molecular structure of the sulfadiazine degradation pathway.

2.2. Design of the experiment

The total length of Qixing Quay in Huixian karst wetland is 1,600 m, and there are large areas of *Eichhornia crassipes* growing in the latter half of the water area of about 300–400 m. The river width is 5.2–51.4 m and mainly ranges from 20 to 30 m, and the water depth is about 0.5–3.0 m. To simulate the actual situation of Qixing Quay, the Huixian karst wetland simulation system was designed (Figure 2). The whole system was outdoors with rain shelter measures. The overall length and width of the system is approximately 100 times smaller than that of the Huixian wetland simulation system. The wetland device specifications were 3.10 m × 1.2 m × 0.4 m, the total length was 12 m, the width of each corridor was 0.3 m, the height was 0.4 m, and the width of the outlet was 0.1 m. The water was directly fed by a peristaltic pump with a hydraulic load of 0.05 m³·(m² d)$^{-1}$ and a water depth of 0.3 m. Different plant groups were separated by baffles. The bottom of the plant group was equipped with a cylindrical fixed device, and many holes conducive to the passage of water were welded at the bottom. The device was used to carry plants for hydroponics. There were eight channels in the river channel of the simulation system, and a partition was set between the blank group and the experimental group. Ten plants per species were hydroponically supplied with Hoagland nutrient solution (Supplementary Table S1) and ultrapure water, and tin foil was used to shield the root area from light. The water inflow mode of the system was continuous water inflow, and the hydraulic retention time was 6 d. The plants in the system underwent an adaptation period of 1 month. If the plant dies during the 1-month adaptation period, then a plant of the same level of growth must be selected for reseeding. Stress culture was conducted after the plants had adapted to the growth conditions.

In the stress culture, according to the water consumption of the simulation system, referring to the results of Xia *et al.*'s (Xia *et al.* 2021) study and combining the detection limits of the actual instrument, SD was added only to the experimental group so that the aqueous solution contained 0.10 μg L^{-1} SD, which was used as a stressor. After 10 d of cultivation, the plants in each group were taken out, and the root exudates of plants were extracted and analyzed by the root soaking method (Zhalnina *et al.* 2018).

The root exudates used were the original aqueous solutions collected from the two wetland plants after a 10-d cultivation under SD stress with a concentration of 0.10 μg L^{-1}. The collected original aqueous solution was filtered with 0.45 μm filter membranes to prepare a culture medium containing 0.10 μg L^{-1} SD. The rhizosphere soil of each plant was collected by the

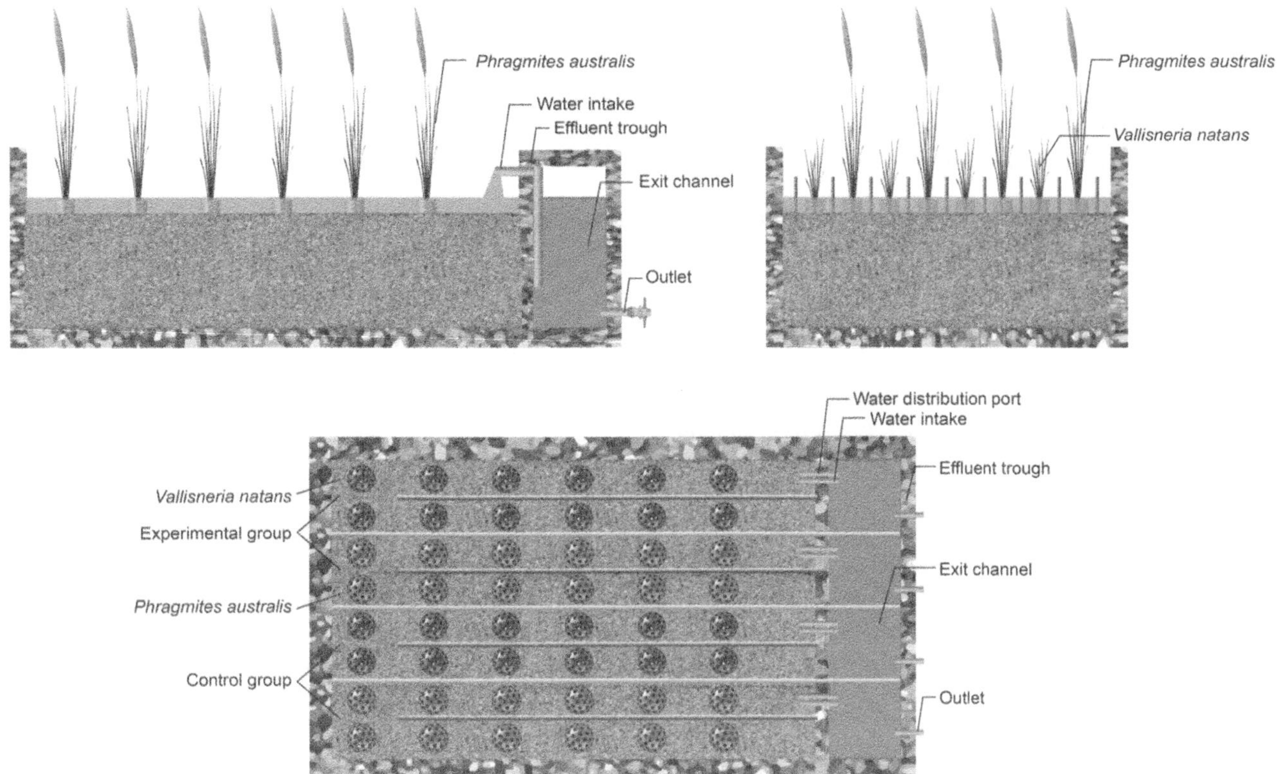

Figure 2 | Simulation of surface flow wetland reactors.

root shaking method (Tu *et al.* 2019). One hundred grams of rhizosphere soil collected from each plant was weighed and placed in a conical flask, and 300 mL of the above antibiotic-containing culture medium was added. The soil and root exudates were fully mixed. After determining that the rhizosphere soil and root exudates of plants corresponded, they were recorded as *P. australis* experimental group (P), *P. australis* control group (CP), *V. natans* experimental group (V), and *V. natans* control group (CV). In the SD degradation experiment, the oscillation time was set at $t = 0/2/8/16/30/48/96$ h as seven time points, and the concentration of SD in the aqueous solution was measured at each time point to obtain the SD degradation rate in different time periods. In the degradation experiment, *P. australis* and *V. natans* were termed as PD and VD, respectively. The rhizosphere soil of 0 and 96 h samples collected to determine microbial community structure, and the aqueous solutions of 0 and 96 h samples were used to detect SD degradation products. The 0 h *P. australis* and *V. natans* groups were labeled as P_0 and V_0, respectively, and the 96 h *P. australis* and *V. natans* groups were P_{96} and V_{96}, respectively.

2.3. Determination of root exudate components

The solution collected by the root soaking method was analyzed by gas chromatography–mass spectrometry (GC-MS) after treatment. The root washing solution was extracted with 300 mL CH_2Cl_2 three times and the CH_2Cl_2 extraction solution was concentrated to dry by rotary evaporation at 35 °C. Five milliliters of CH_2Cl_2 filtered by a 0.45-μm membrane was then added, the solution was dried with anhydrous Na_2SO_4, and finally, 0.5 mL of treated CH_2Cl_2 was taken for GC-MS analysis. The instrument used for determination was a GC-MS-type gas chromatography–mass spectrometer (Claus 600T Mass Spectrometer and Claus 680 Gas Chromatograph), Elite-5MS (30 m × 0.25 mm × 0.25 μm) chromatographic column.

2.4. Determination of SD content and degradation products

SD was detected by solid phase extraction-high-performance liquid chromatography. The instruments used included the following: a Japan AQUAT race ASPE799 solid phase extraction instrument, Agilent 1260 high-performance liquid chromatography, a 24-bit nitrogen blowing instrument, an HLB extraction column, and an ultrasonic cleaner.

The detection samples of SD degradation products were tested Guangxi Guilin RID Testing Co., Ltd. A triple quadrupole liquid chromatography–mass spectrometry (HPLC-MS) system was used with a chromatographic column: an Eclipse Plus C18 RRHD 2.1 × 50 mm, 1.8 μm. Sample treatment was conducted as follows: the sample solution was moved to a centrifuge tube, centrifuged at 4,000 r min^{-1} for 5 min, and the supernatant was collected for filtration. One hundred milliliters of the sample solution were taken in a conical flask with a measuring tube. The sample solution was concentrated to 100 times by solid phase extraction and detected by HPLC-MS/MS.

2.5. Bacterial DNA extraction and PCR amplification from rhizosphere soil

The E. Z. N. ATM Mag-Bind Soil DNA Kit (Omega Bio-Tek Company, Guangzhou International Business Incubator) was used for DNA extraction, and then the Qubit 3.0 DNA detection kit was used for the accurate quantification of genomic DNA to determine the amount of DNA added to the PCR reaction. The PCR primers were fused with 16S V3–V4 primers on the sequencing platform. The PCR reaction conditions were 94 °C, 3 min → (94 °C, 30 s → 45 °C, 20 s → 65 °C, 30 s) 5 → (94 °C, 20 s → 55 °C, 20 s → 72 °C, 30 s) 20 →72 °C, 5 min → 10 °C. DNA extraction, PCR amplification and purification were performed by the Sangon Biotech Co., Ltd (Shanghai).

2.6. Statistical analysis

UPARSE software was used for operational taxonomic unit (OTU) sequence clustering of valid sample data. The Mothur method and SILVA's SSU rRNA database were used to perform species annotation analysis on the representative OTU sequences and obtain the community composition of the sample at each taxonomic level. The alpha diversity index was calculated using Qiime 1.9.1. A species abundance column diagram, degradation rate analysis diagram, degradation kinetics diagram, and dilution curve were analyzed and drawn by Origin 2018. The *p*-value was obtained using Welch's *t*-test. Abundance differences were plotted using STAMP software. Data analysis of the results was performed in SPSS 25.0.

3. RESULTS AND DISCUSSION

3.1. Composition of root exudates

Each measured peak (Supplementary Figure S1(a)–S1(d)) was searched in the atlas library to obtain the corresponding compounds (Supplementary Table S2). A total of 37 compounds were detected in the CP group, including alkanes, aldehydes, esters, alcohols, and amides. A total of 31 compounds were detected in group P, including alkanes and esters. A total of 35 compounds were detected in the CV group, including alkanes, aldehydes, esters, amides, and alcohols. A total of 34 compounds were detected in root exudates of group V, including alkanes and esters.

From the perspective of the most abundant component (Table 1), the most abundant component of the CP group was heptadecane, 2,6,10,15-tetramethyl, with a content of 8.46%, but that of the P group was 9-octadecenoic acid (Z)-, methyl ester, with a content of 31.88%. In the CV group, the content of 9-octadecenamide, (Z)- was 9.6%, and in the V group, the content of 9-octadecenoic acid (Z)-, methyl ester was 22.31%. According to the above results, the most abundant components of the two plant groups changed to esters under SD stress, and the content increased. Under material stress, the secretion of some compounds in plant root exudates was inhibited or changed (Rolfe *et al.* 2019). In this study, dichloromethane was used for extraction, and the obtained experimental results covered relatively complete root exudates.

Studies used different nitrogen and phosphorus concentrations to culture plants, and they concluded that the amount of compounds in the root exudates of the low nutrient treatment was higher than that of the high nutrient treatment (Wu *et al.* 2012). However, the detection of compounds in this study was higher than that in the above studies. Some scholars have found that the root exudates of rape seedlings mainly consist of compounds including hydrocarbons, alcohols, esters,

Table 1 | Changes in the most abundant components in rhizosphere secretions before and after degradation

Plant group	Substance name	Category	Content (%)
CP	Heptadecane, 2,6,10,15-tetramethyl	Alkanes	8.46
P	9-Octadecenoic acid (Z)-, methyl ester	Esters	31.88
CV	9-Octadecenamide (Z)-	Amides	9.6
V	9-Octadecenoic acid (Z)-, methyl ester	Esters	22.31

and acids (Escolà Casas & Matamoros 2021). This finding is similar to the results in this study, but the specific characteristics of compounds are not consistent. After SD stress at a concentration of 0.10 µg L^{-1}, the species of compounds in the rhizosphere exudates of the two plants decreased, which may have been due to the fact that wetland plants would actively adapt to the environment by self-regulating the composition and quantity of secretions by roots under external stress (Duan *et al.* 2020).

3.2. Degradation analysis of SD

As shown in Figure 3, the degradation efficiency of antibiotics in all of the plant groups increased with time, and was the highest at 96 h and the lowest at 2 h, the degradation efficiency of VD was 15.53% in 2 h, and the degradation efficiency of PD reached 12.40% in 2 h. From 2 to 30 h, the degradation efficiency of VD was always higher than that of PD. From 30 to 96 h, the degradation rate of PD gradually became higher than that of VD, the degradation rate of VD was 1.15, 0.98, and 0.57 µg L^{-1} h^{-1} at 30, 48, and 96 h, and the degradation rate of PD reached 0.74, 0.65, and 0.55 µg L^{-1} h^{-1} at the same time. At 96 h, the degradation efficiency of the *P. australis* group was 97.20% (Supplementary Table S4), and that of the *V. natans* group was 85.60% (Supplementary Table S4).

The experimental results showed that the removal rate of antibiotics with these species was higher than that of other constructed wetlands or non-karst wetland plants (Yan *et al.* 2019). Owing to the long-term erosion of carbonate rocks, the Huixian wetland ecosystem contains a large amount of Ca^{2+} (Li *et al.* 2017), and the form of calcium that occurs in the soil used in this study is mainly exchangeable, with an ECa/TCa value of 52.82–69.48% (Supplementary Table S3). The high content of exchangeable calcium indicates that calcium in the soil is active in migration and bioavailability, and Ca^{2+} can form a complex with antibiotics and then be adsorbed by the soil matrix (Liang *et al.* 2018). However, SAs are a typical amphoteric compound. When the soil is alkaline, antibiotics are difficult to adsorb, and when the pH is close to neutral, the removal effect of antibiotics is the best (Kurade *et al.* 2019). Huixian wetland soil is rich in calcium and alkaline, and its water quality is mostly weak alkaline. In this study, the rhizosphere soil pH was 7.16–7.22 (Supplementary Table S3), which indicated that SD was difficult to adsorb by the soil matrix in the karst wetland, and probably existed in the form of complexes. In addition, the removal mechanism of SD in Huixian wetland may have been mainly dependent on the interaction between plants and rhizosphere microorganisms.

The data were fitted by the first-order equation and the second-order equation of degradation kinetics (Figure 4(a) and 4(b)). The residual concentration of SD decreased gradually with time, indicating that the degradation effect increased with time, which was consistent with the expression of the degradation rate. Figure 4(a) and 4(b) and Supplementary Table S4 also show that the remaining concentration gradually slows down after 48 h and tends to flatten at 96 h. In wetland environments, the plant rhizosphere is often flooded. The soil carbon mineralization rate under anaerobic conditions is much lower than that

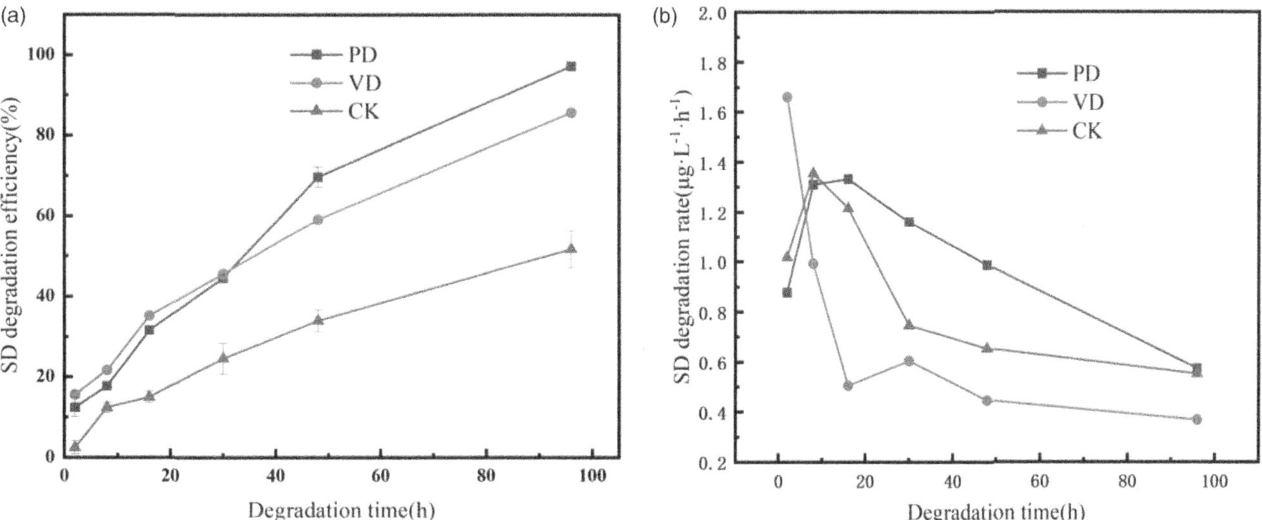

Figure 3 | Variation of sulfadiazine (SD) degradation efficiency (a) and degradation rate with time (b).

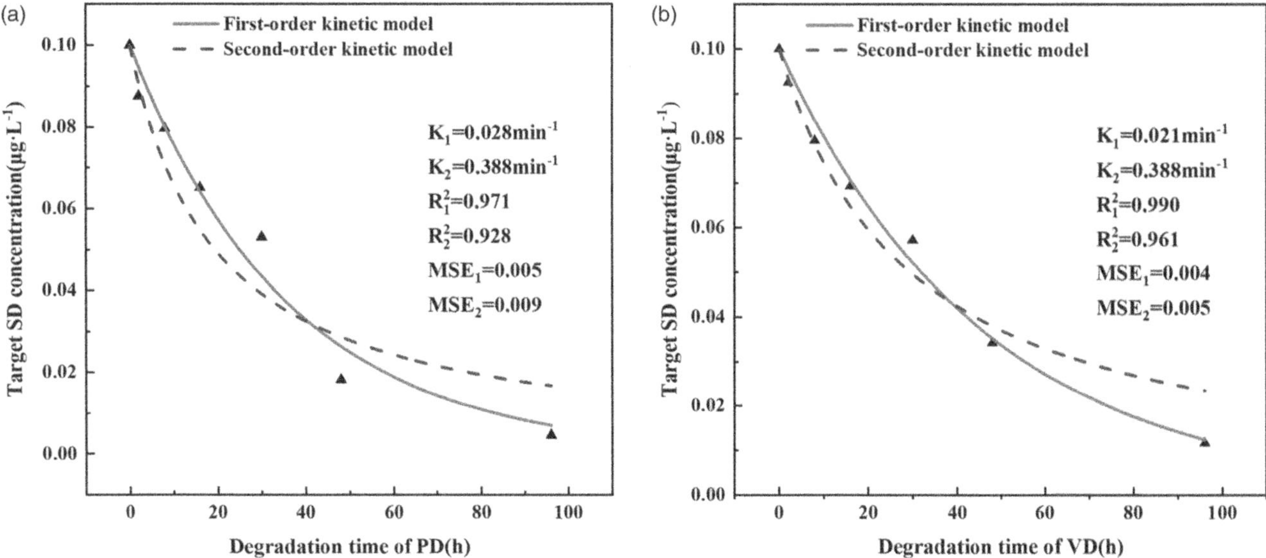

Figure 4 | SD degradation kinetics of *P. australis* (PD) (a) and *V. natans* (VD) (b); K_1 and K_2 are the first- and second-order kinetic degradation rate constants, respectively, R_1^2 and R_2^2 are the first- and second-order kinetic fitting coefficients, respectively, and MSE_1 and MSE_2 are the mean square deviation of the target concentration expressed by the first- and second-order kinetics, respectively.

under aerobic conditions, resulting in high organic carbon content in wetlands (Dos Santos Teixeira *et al.* 2021). Supplementary Table S3 also shows that the wetlands had abundant carbon sources (21.86–22.28 g kg^{-1}). SD can be used as a microbial carbon source, but a rich carbon source in soil will reduce microbial demand for SD. With the decline of a biodegradable carbon source in the environment, microorganisms will use SD as a carbon source, and the biodegradation of SD will be enhanced. However, with the continuous consumption of SD, microbial activity will gradually decrease.

The R_1^2 values of PD and VD were higher than R_2^2 and MSE_1 was lower than MSE_2, indicating that the fitting degree and correlation of the first-order equation of degradation kinetics were higher than those of the second-order equation, and the first-order kinetic model was more reasonable to express the degradation of SD in this study. This result was consistent with the results of research on the photodegradation, Fenton-oxidation degradation, and chlorination degradation of antibiotics (Dirany *et al.* 2010). Different plants had different pollutant removal capacities, and the selection of plant species was a key part of pollutant removal (Rezania *et al.* 2015). Therefore, the cultivation of *P. australis* was of great significance for the prevention and control of SD in Huixian wetland.

3.3. Analysis of SD degradation products

Microorganisms can degrade the phenyl portion of SD molecules, but the pyrimidine ring of SD is stable and produces an equal molar amount of 2-aminopyrimidine. When 2-aminopyrimidine is formed, it will be transformed into 4-hydroxy-2-aminopyrimidine by bacterial metabolism (Deng *et al.* 2016). The intermediates of SD also include sulfonamide aniline acid, p-aminobenzenesulfonic acid, 2-hydroxypyrimidine, aniline, and pyrimidine-2-sulfonamide acid. The degradation pathways of the intermediates may be one of three parallel pathways: first, the S–N bond, N–C bond, or C–S bond is broken, and then hydroxylation, formylation, and acetylation occur (Mohring *et al.* 2009). In this experiment, the two plant groups obtained the same three metabolites (Supplementary Figure S2(a) and S2(b)): *m/z* 171.2 (p-aminobenzenesulfonic acid), *m/z* 114.1 (4-hydroxy-2-aminopyridine), and *m/z* 224.2 (phenyl sulfoxide).

Therefore, we can deduce the possible degradation pathway of SD (Figure 5). I indicates that the product p-aminobenzenesulfonic acid was obtained by hydrolysis of the S–N bond and hydroxylation; II indicates that the product 4-hydroxy-2-aminopyridine was obtained by hydrolysis of the S–N bond, deamination of the C atom on the connecting chain, and then amination and hydroxylation; III shows the hydrolysis, hydroxylation, and deamination of the S–N bond to obtain phenyl sulfoxide. The results are consistent with previous studies; that is, SAs can be degraded by chlorine substitution, S–N bond cleavage, S–C bond cleavage, hydroxylation/oxidation, and desulfonation, and they can generate a series of degradation products (Wang & Helbling 2016).

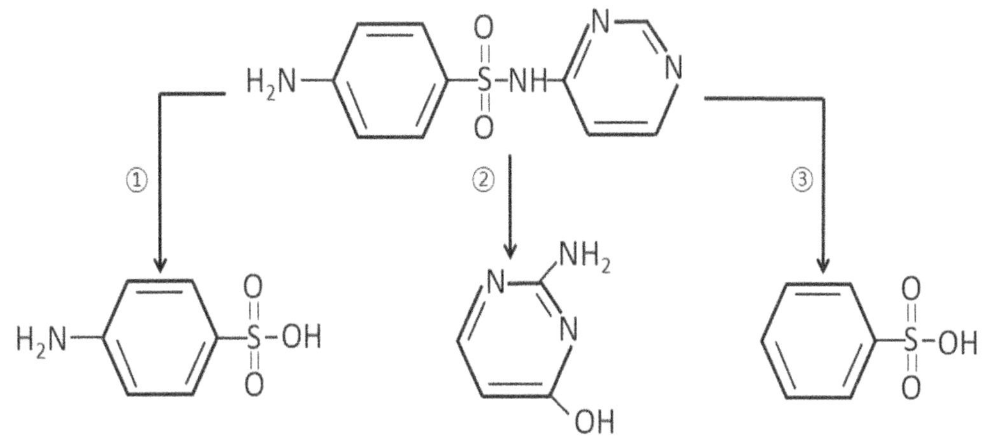

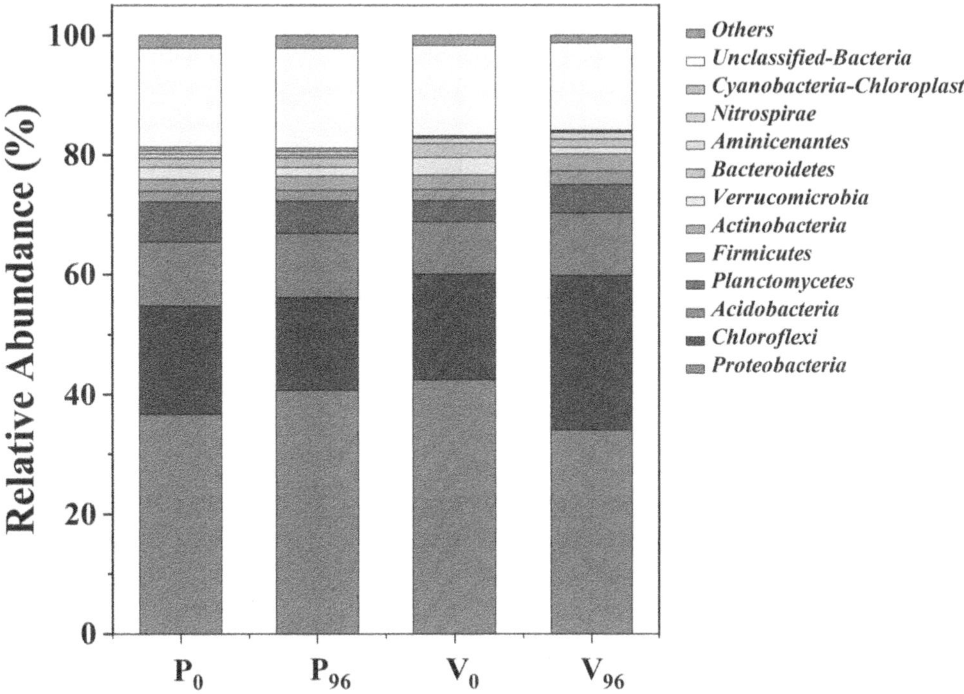

Figure 5 | Possible degradation pathways of sulfadiazine (SD).

3.4. Effect of SD degradation on rhizosphere microorganism phyla and genus abundance

3.4.1. The relative division of microorganisms at the phylum level

In this research, 37 phyla, 87 classes, 137 orders, 238 families, and 448 genera were detected, indicating that the bacteria were widely distributed, and the bacterial community had high diversity. Thirteen categories with the highest abundances were selected to generate histograms (Figure 6), including *Proteobacteria* (34.49–43.21%), *Chloroflexi* (15.75–26.11%), *Acidobacteria* (8.92–11.02%), *Planctomycetes* (3.58–6.77%), *Firmicutes* (1.79–2.31%), *Actinobacteria* (2.01–2.85%), *Verrucomicrobia* (1.11–3.12%), *Bacteroidetes* (1.41–2.41%), *Aminicenante* (0.46–1.06%), *Nitrospirae* (0.19–0.66%), and *Cyanobacteria-Chloroplast* (0.14–0.54%) (Supplementary Table S5), which were similar to those reported in other rhizosphere bacterial communities (Hua *et al.* 2018; Tian *et al.* 2020; Yang *et al.* 2021). Tu *et al.* (2019) also investigated the distribution characteristics of rhizosphere bacterial communities in Huixian wetland, but the bacterial abundance in this study was lower than

Figure 6 | Comparison of relative abundance at the phylum level.

that in their research, indicating that the presence of SD hindered the growth of certain bacterial communities. Bacteria that degrade antibiotics are mainly composed of *Proteobacteria*, *Bacteroidetes*, *Acidobacteria*, *Actinobacteria*, *Verrucomicrobia*, and *Phytoplankton* (Wu *et al.* 2019). The relative abundances of *Proteobacteria*, *Acidobacteria*, *Actinobacteria*, and *Bacteroidetes* in rhizosphere bacterial communities in this study were high, which was one of the reasons for the high degradation rate in the two plant groups. The dominant phyla did not change before and after degradation, but their relative abundances changed to varying degrees.

3.4.2. The relative division of microorganisms at the phylum level

The first 34 genera with the highest abundances were selected to generate a relative abundance histogram (Figure 7). All of the samples were composed of 34 genera, of which 20 were identified, and the unclassified genera accounted for 42.98–52.17% (Supplementary Table S6). Before and after degradation, the abundance of dominant bacteria in different plant groups changed to varying degrees. It is worth noting that in the *P. australis* group, new genera appeared after 96 h of SD degradation, which was *Methylophilus*, and they belonged to *Methylophilaceae* at the upper level. *Methylophilaceae*, as a kind of *Proteobacteria*, can increase with the concentration of antibiotics in soil-lettuce systems (Shen *et al.* 2021). This is due to *Methylophilaceae*'s potential to degrade exogenous chemicals (Shen *et al.* 2019).

Root exudates can be transported by cells and excreted around the rhizosphere, creating a unique environment for root microorganisms (Hu *et al.* 2018), thus affecting the degradation of SAs. This work found that after SD stress culture, the highest component of root exudates was changed to organic acid esters. In the presence of acid or alkali, organic acid esters can be hydrolyzed to organic acids or alcohols. Many studies have shown that plant roots secrete organic acids, amino acids, sugars, and other substances that contribute to microbial growth (Li *et al.* 2019). Biodegradation has been proved to be the main pathway for SA degradation in plants (Chen & Xie 2018). The contribution of plant roots and pollutants to microbial growth is different. Plant root exudates can increase the activity of microorganisms, and some microorganisms can promote the activity of plants, helping them resist pollutant stress (Cristaldi *et al.* 2017). Therefore, the calcium-rich and alkali-rich characteristics of karst wetland soil will lead to the hydrolysis of some organic acid esters in root exudates

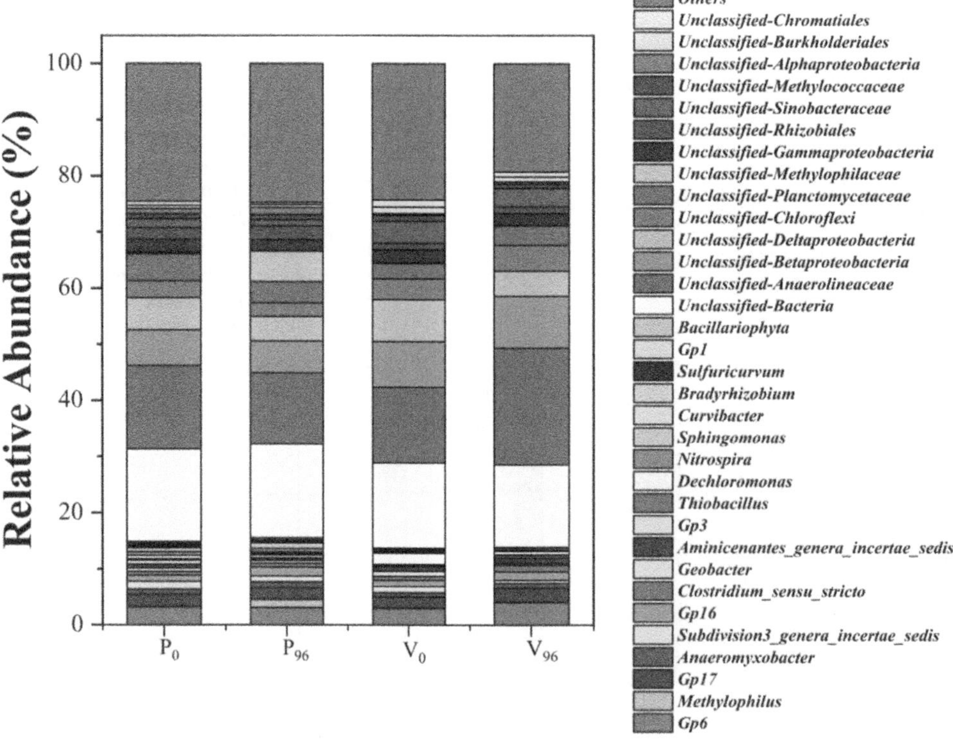

Figure 7 | Comparison of relative abundance at the genus level.

to produce organic acids, which indirectly affects the rhizosphere microbial metabolism, and thus regulates the degradation of antibiotics.

3.5. Difference analysis of bacterial community abundance before and after decomposition

Figure 8 shows the proportion of microbial abundance differences before and after decomposition in the 95% confidence interval. The p-value is shown on the rightmost side. This graph only lists the 25 lowest p-values. At the phylum level, the abundance of *Synergistetes* increased significantly after degradation.

At the genus level (Figure 9), there were 25 genera with significant differences before and after degradation ($p < 0.05$), among which 6 genera were extremely significant ($p < 0.01$). These genera were *Piscinibacter, Unclassified-Xanthomonadaceae, Methylotenera, Streptococcus, Unclassified-Methylophilaceae*, and *Nitrosomonas*. They were all significantly increased compared with those before degradation, indicating that SD degradation played a catalytic role in them.

Root exudates have been shown to affect the composition of rhizosphere microbial communities. For example, salicylic acid can induce systemic resistance in plants and inhibit the growth of pathogens (Badri *et al.* 2013), benzoic acid in

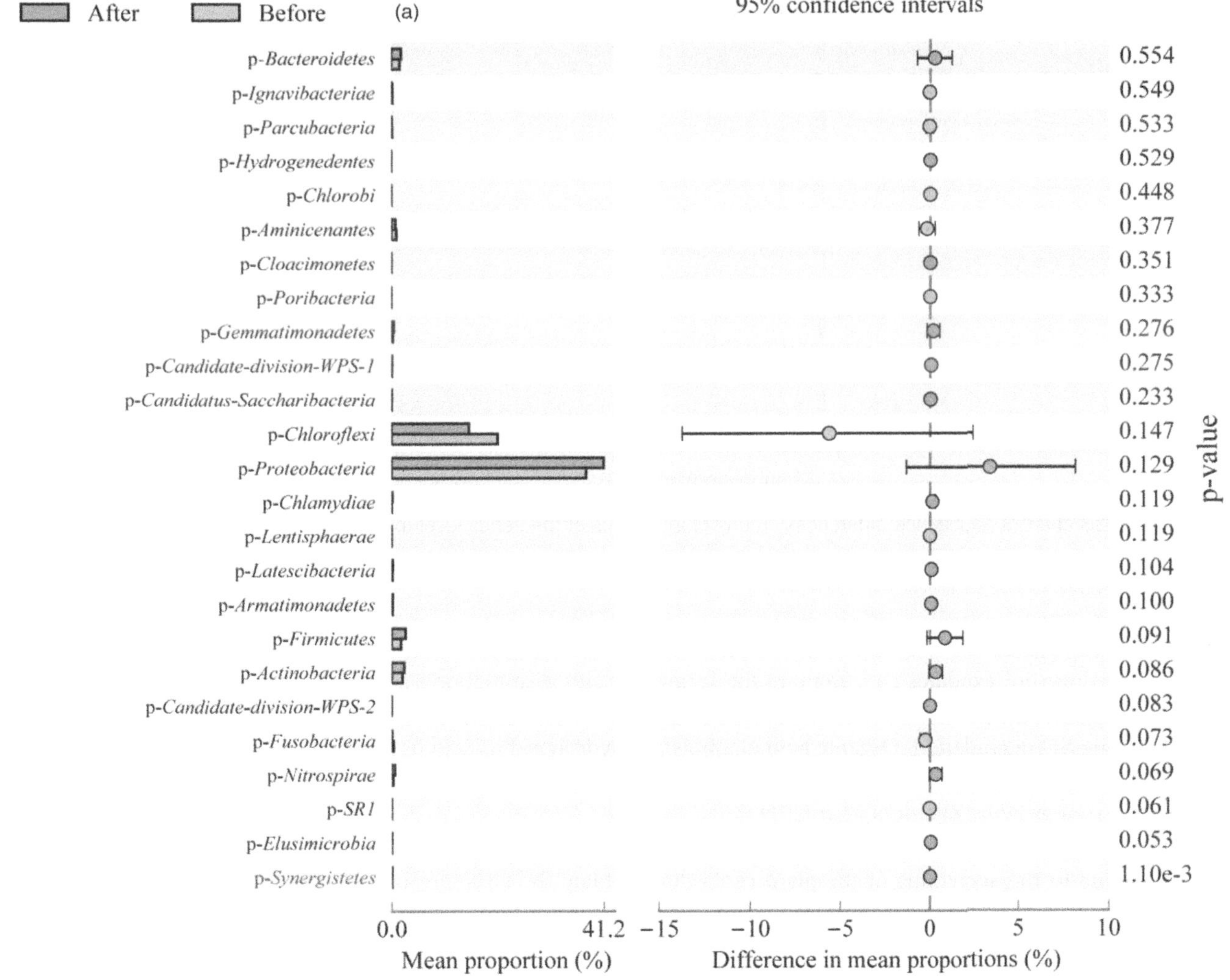

Figure 8 | Analysis chart of abundance differences among plant groups at the phylum level (a); 'Before' indicates the relative abundance of rhizosphere microorganisms when the plant group had degraded SD for 0 h, while 'After' indicates the relative abundance of rhizosphere microorganisms when the plant group had degraded SD for 96 h. Red text represents the species with significant or extremely significant differences before and after degradation (the same below). Please refer to the online version of this paper to see this figure in colour: https://dx.doi.org/10.2166/wrd.2023.062.

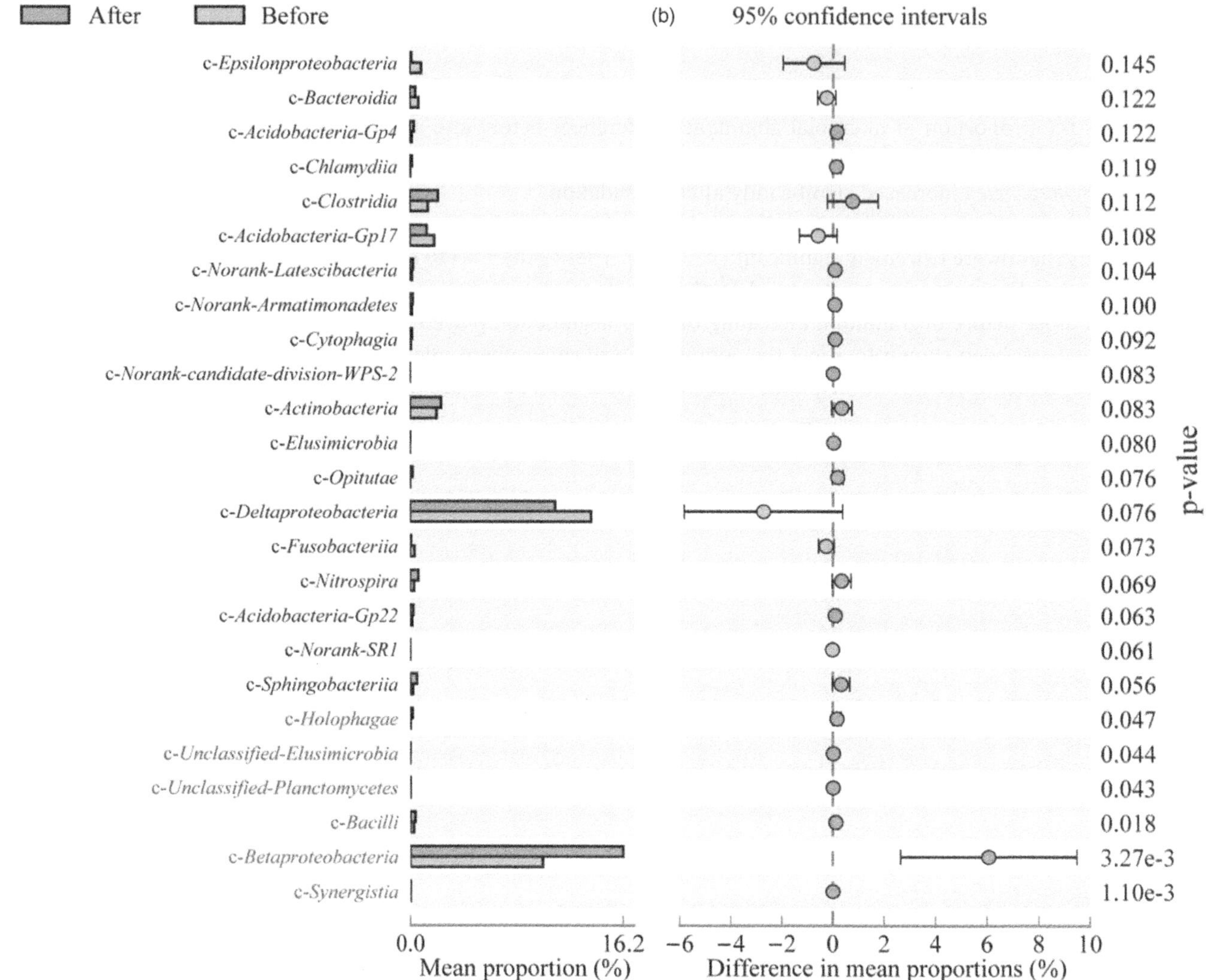

Figure 9 | Analysis chart of abundance differences among plant groups at the genus level (b).

peanut root exudates increases the relative abundance of *Burkholderiaspp* in rhizosphere soil (Liu *et al.* 2017), and ferulic acid in watermelon root exudates can promote the formation and germination of *Fusarium oxysporum* spores (Hao *et al.* 2010). In this experiment, it has been shown that a large amount of organic acid alcohols is produced in the rhizosphere secretions during SD degradation. Organic acid alcohols are hydrolyzed in acid or alkaline environments to produce corresponding acids and alcohols. Some strains exhibit significant absorption of amino acids, organic acids, sugars, and quaternary amines during root growth (Zhalnina *et al.* 2018).

The selective influence of plants on the composition and structure of the inter-rooted microbial community is always dominant, mainly due to the secretions of the plant roots that regulate the structure of the inter-rooted microbial community, which differs in root exudates before and after SD degradation. These root secretions can induce and stimulate the growth of specific bacterial groups and influence the abundance and diversity of inter-rooted microorganisms, which in turn has an impact on degradation efficiency and products. In a study of SD degradation dynamics, it was found that the areas with the highest degradation rates were those with the highest microbial abundance, and that the community structure changed before and after the experiment, probably because the crop roots secreted large amounts of organic matter into the soil, and these root secretions in turn directed the microbial community toward reducing external stresses, promoting SD degradation, causing the inter-rooted microbial community structure to respond.

3.6. Diversity analysis of bacterial communities

The curve was drawn with the alpha diversity index as the vertical axis (Figure 10). All of the curves were not stable; that is, each sample did not reach saturation, indicating that the diversity of rhizosphere bacteria in all of the plant groups in this study was high. When the sequencing quantity is more than 20,000, the curve growth trend slows down until it tends to flatten, indicating that the sample sequencing quantity is reasonable. The diversity and abundance of bacteria changed before and after the degradation of SD. In addition, the OTU number of the *P. australis* group was higher than that of the *V. natans* group.

As shown in Table 2, the number of OTUs in the samples was 2,541–3,353, and the coverage was over 98% (coverage: 98–99%). It can be seen from the table that for the Shannon $P_0 > P_{96} > V_0 > V_{96}$ and for the Simpson $V_{96} > V_0 > P_{96} > P_0$; thus, the bacterial diversity of the rhizosphere in the *P. australis* group was always higher than that in the *V. natans* group before degradation (0 h) and after degradation (96 h). In the above analysis of the degradation rate, that of the *P. australis* group was also greater than that of the *V. natans* group, which was consistent with the expression of diversity analysis. The main bacterial groups involved in SA degradation may be resistant to antibiotics, thus contributing to SA biodegradation (Yang *et al.* 2016).

It can also be found from the table that the rhizosphere bacterial diversity of *P. australis* and *V. natans* decreased after 96 h of SD degradation. Chao1 was $P_{96} > P_0 > V_{96} > V_0$ and Shannoneven was $P_0 > P_{96} > V_0 > V_{96}$, indicating that the degradation of SD increased the total number of bacteria in the rhizosphere of plants, but decreased the uniformity of bacterial distribution. Wang *et al.* (2021b) found that adding root exudates reduced microbial community diversity, but increased community abundance. This may have been due to the presence of a large amount of organic matter in plant root exudates. This

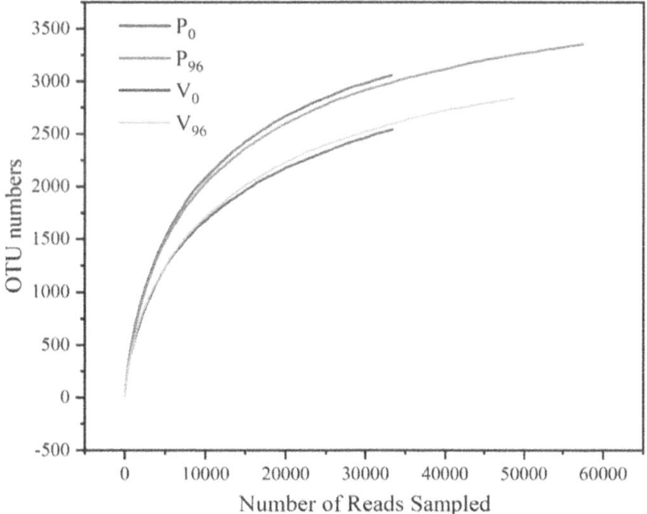

Figure 10 | Dilution curves of plant groups. P_0 was the group of *P. australis* rhizosphere degradation for 0 h, P_{96} was the group of *P. australis* rhizosphere degradation for 96 h, V_0 was the group of *V. natans* rhizosphere degradation for 0 h, and V_{96} was the group of *V. natans* rhizosphere degradation for 0 h.

Table 2 | Alpha diversity index of each plant group

Sample	Shannon	OTUs	Chao	Simpson	Shannoneven	Coverage (%)
V_0	6.39	2,541	3,081.98	0.005	0.82	98
V_{96}	6.32	2,838	3,328.27	0.007	0.80	99
P_0	6.89	3,055	3,509.62	0.004	0.86	98
P_{96}	6.78	3,353	3,678.60	0.006	0.83	99

leads to the evolution of the microbial community toward the reduction of external stress, which promotes the degradation of SD and induces and stimulates the growth of specific bacterial communities, thereby affecting the abundance and diversity of rhizosphere microorganisms (Yuan *et al.* 2017).

4. CONCLUSION

In order to reveal the sulfonamide degradation mechanism of the wetland plant rhizosphere, we compared the composition of rhizosphere exudates of *P. australis* and *V. natans* under 0.10 μg L^{-1} SD stress in this study. The degradation of SD by rhizosphere exudates extracted under stress was also observed. We found that SD stress could increase the content of organic acid esters in plant roots. In the degradation experiment, we found that the degradation effect of the *P. australis* group was stronger than that of the *V. natans* group. During the degradation process, the relative abundance of rhizosphere bacteria will change, the diversity will decrease, and new genera will be introduced. In addition, we conclude that the degradation of SD in plants in karst areas may be controlled indirectly by increasing the content of organic acid esters in roots. Under weak alkaline conditions, organic acid esters break down to produce organic acids that affect the rhizosphere microbial metabolism.

FUNDING

This research was funded by the Guangxi Natural Science Foundation (grant number 2022GXNSFFA035033), the National Natural Science Foundation of China (grant number 52260023 and 51878197); the Basic Ability Enhancement Program for Young and Middle-aged Teachers of Guangxi (grant number 2021KY0265); and Innovation Project of Guangxi Graduate Education (YCBZ2022117).

PUBLISHER'S NOTE

All claims expressed in this article are solely those of the authors and do not necessarily represent those of their affiliated organizations, or those of the publisher, the editors and the reviewers. Any product that may be evaluated in this article, or claim that may be made by its manufacturer, is not guaranteed or endorsed by the publisher.

DATA AVAILABILITY STATEMENT

All relevant data are included in the paper or its Supplementary Information.

CONFLICT OF INTEREST

The authors declare there is no conflict.

REFERENCES

Badri, D. V., Chaparro, J. M., Zhang, R., Shen, Q. & Vivanco, J. M. 2013 Application of natural blends of phytochemicals derived from the root exudates of Arabidopsis to the soil reveal that phenolic-related compounds predominantly modulate the soil microbiome. *Journal of Biological Chemistry* **288** (7), 4502–4512.

Chen, J. & Xie, S. 2018 Overview of sulfonamide biodegradation and the relevant pathways and microorganisms. *Science of the Total Environment* **640**, 1465–1477.

Chen, J., Wei, X.-D., Liu, Y.-S., Ying, G.-G., Liu, S.-S., He, L.-Y., Su, H.-C., Hu, L.-X., Chen, F. & Yang, Y.-Q. 2016 Removal of antibiotics and antibiotic resistance genes from domestic sewage by constructed wetlands: optimization of wetland substrates and hydraulic loading. *Science of the Total Environment* **565**, 240–248.

Chen, J., Liu, S.-S., He, L., Cheng, Y.-X., Ye, P., Li, J., Ying, G., Wang, Y.-J. & Yang, F. 2021 The fate of sulfonamides in the process of phytoremediation in hydroponics. *Water Research* **198**, 117145.

Conkle J, L., White J, R. & Metcalfe C, D. 2008 Reduction of pharmaceutically active compounds by a Lagoon wetland wastewater treatment system in Southeast Louisiana. *Chemosphere* **73** (11), 1741–1748.

Cristaldi, A., Conti, G. O., Jho, E. H., Zuccarello, P., Grasso, A., Copat, C. & Ferrante, M. 2017 Phytoremediation of contaminated soils by heavy metals and PAHs: a brief review. *Environmental Technology & Innovation* **8**, 309–326.

Cui, J., Fu, L., Tang, B., Bin, L., Li, P., Huang, S. & Fu, F. 2020 Occurrence, ecotoxicological risks of sulfonamides and their acetylated metabolites in the typical wastewater treatment plants and receiving rivers at the Pearl River Delta. *Science of the Total Environment* **709**, 136192.

Deng, Y., Mao, Y., Li, B., Yang, C. & Zhang, T. 2016 Aerobic degradation of sulfadiazine by *Arthrobacter* spp.: kinetics, pathways and genomic characterization. *Environmental Science & Technology* **50** (17), 9566–9575.

Dirany, A., Sirés, I., Oturan, N. & Oturan, M. A. 2010 Electrochemical abatement of the antibiotic sulfamethoxazole from water. *Chemosphere* **81** (5), 594–602.

Dodgen, L., Kelly, W., Panno, S., Taylor, S., Armstrong, D., Wiles, K., Zhang, Y. & Zheng, W. 2017 Characterizing pharmaceutical, personal care product, and hormone contamination in a karst aquifer of southwestern Illinois, USA, using water quality and stream flow parameters. *Science of the Total Environment* **578**, 281–289.

Dos Santos Teixeira, A. F., Silva, S. H. G., Soares de Carvalho, T., Silva, A. O., Azarias Guimarães, A. & de Souza Moreira, F. M. 2021 Soil physicochemical properties and terrain information predict soil enzymes activity in phytophysiognomies of the Quadrilátero Ferrífero region in Brazil. *CATENA* **199**, 105083.

Du, L., Zhao, Y., Wang, C., Zhang, H., Chen, Q., Zhang, X., Zhang, L., Wu, J., Wu, Z. & Zhou, Q. 2020 Removal performance of antibiotics and antibiotic resistance genes in swine wastewater by integrated vertical-flow constructed wetlands with zeolite substrate. *Science of the Total Environment* **721**, 137765.

Duan, X., Zhao, Y.-Y. & Zhang, J.-C. 2020 Characteristics of the root exudate release system of typical plants in plateau lakeside wetland under phosphorus stress conditions. *Open Chemistry* **18** (1), 808–821.

Escolà Casas, M. & Matamoros, V. 2021 Analytical challenges and solutions for performing metabolomic analysis of root exudates. *Trends in Environmental Analytical Chemistry* **31**, e00130.

Hao, W.-Y., Ren, L.-X., Ran, W. & Shen, Q.-R. 2010 Allelopathic effects of root exudates from watermelon and rice plants on *Fusarium oxysporum* f.sp. *niveum*. *Plant and Soil* **336** (1), 485–497.

He, L.-Y., Ying, G., Liu, Y.-S., Su, H.-C., Chen, J., Liu, S.-S. & Zhao, J.-L. 2016 Discharge of swine wastes risks water quality and food safety: antibiotics and antibiotic resistance genes from swine sources to the receiving environments. *Environment International* **92–93**, 210–219.

Hillebrand, O., Nödler, K., Sauter, M. & Licha, T. 2015 Multitracer experiment to evaluate the attenuation of selected organic micropollutants in a karst aquifer. *Science of the Total Environment* **506–507**, 338–343.

Hu, L., Robert, C. A. M., Cadot, S., Zhang, X., Ye, M., Li, B., Manzo, D., Chervet, N., Steinger, T., van der Heijden, M. G. A., Schlaeppi, K. & Erb, M. 2018 Root exudate metabolites drive plant-soil feedbacks on growth and defense by shaping the rhizosphere microbiota. *Nature Communications* **9** (1), 2738.

Hu, H., Zhou, Q., Li, X., Lou, W., Du, C., Teng, Q., Zhang, D., Liu, H., Zhong, Y. & Yang, C. 2019 Phytoremediation of anaerobically digested swine wastewater contaminated by oxytetracycline via *Lemna aequinoctialis*: nutrient removal, growth characteristics and degradation pathways. *Bioresource Technology* **291**, 121853.

Hua, G., Cheng, Y., Kong, J., Li, M. & Zhao, Z. 2018 High-throughput sequencing analysis of bacterial community spatiotemporal distribution in response to clogging in vertical flow constructed wetlands. *Bioresource Technology* **248**, 104–112.

Kurade, M. B., Xiong, J.-Q., Govindwar, S., Roh, H.-S., Saratale, G., Jeon, B.-H. & Lim, H. 2019 Uptake and biodegradation of emerging contaminant sulfamethoxazole from aqueous phase using *Ipomoea aquatica*. *Chemosphere* **225**, 696–704.

Lambertini, C., Guo, W.-Y., Ye, S., Eller, F., Guo, X., Li, X.-Z., Sorrell, B. K., Speranza, M. & Brix, H. 2020 Phylogenetic diversity shapes salt tolerance in *Phragmites australis* estuarine populations in East China. *Scientific Reports* **10** (1), 17645.

Li, H.-L., Wang, Y.-Y., Qian, Z., Wang, P., Zhang, M.-X., Yu, F.-H. & Jian, L. 2015 Vegetative propagule pressure and water depth affect biomass and evenness of submerged macrophyte communities. *PLoS ONE* **10** (11), e0142586.

Li, Z., Jin, Z. & Li, Q. 2017 Changes in land use and their effects on soil properties in Huixian karst wetland system. *Polish Journal of Environmental Studies* **26** (2), 699–707.

Li, C.-H., Che, X.-W., Bai, Y., Shi, X.-Y. & Su, R.-G. 2019 Indirect photodegradation of sulfamethoxazole in water. *Environmental Science* **40** (1), 273–280.

Liang, P., Wu, S., Zhang, C., Xu, J., Christie, P., Zhang, J. & Cao, Y. 2018 The role of antibiotics in mercury methylation in marine sediments. *Journal of Hazardous Materials* **360**, 1–5.

Liu, J., Li, X., Jia, Z., Zhang, T. & Wang, X. 2017 Effect of benzoic acid on soil microbial communities associated with soilborne peanut diseases. *Applied Soil Ecology* **110**, 34–42.

Man, Y., Wang, J., Tam, N. F.-y., Wan, X., Huang, W., Zheng, Y., Tang, J., Tao, R. & Yang, Y. 2019 Responses of rhizosphere and bulk substrate microbiome to wastewater-borne sulfonamides in constructed wetlands with different plant species. *Science of the Total Environment* **706**, 135955.

Menció, A. & Mas-Pla, J. 2019 Assessing the influence of environmental factors on groundwater antibiotic occurrence by means of variation partitioning. *Water* **11**, 1495.

Mohring, S. A. I., Strzysch, I., Fernandes, M. R., Kiffmeyer, T. K., Tuerk, J. & Hamscher, G. 2009 Degradation and elimination of various sulfonamides during anaerobic fermentation: a promising step on the way to sustainable pharmacy? *Environmental Science & Technology* **43** (7), 2569–2574.

Qin, L., Pang, X., Zeng, H.-H., Liang, Y.-P., Mo, L.-Y., Wang, D.-Q. & Dai, J.-F. 2019 Ecological and human health risk of sulfonamides in surface water and groundwater of Huixian karst wetland in Guilin, China. *Science of the Total Environment* **708**, 134552.

Rezania, S., Ponraj, M., Talaiekhozani, A., Mohamad, S. E., Din, M. F. M., Taib, S. M., Sabbagh, F. & Sairan, F. M. 2015 Perspectives of phytoremediation using water hyacinth for removal of heavy metals, organic and inorganic pollutants in wastewater. *Journal of Environmental Management* **163**, 125–133.

Rolfe, S., Griffiths, J. & Ton, J. 2019 Crying out for help with root exudates: adaptive mechanisms by which stressed plants assemble health-promoting soil microbiomes. *Current Opinion in Microbiology* **49**, 73–82.

Shen, Y., Stedtfeld, R. D., Guo, X., Bhalsod, G. D., Jeon, S., Tiedje, J. M., Li, H. & Zhang, W. 2019 Pharmaceutical exposure changed antibiotic resistance genes and bacterial communities in soil-surface- and overhead-irrigated greenhouse lettuce. *Environment International* **131**, 105031.

Shen, Y., Ryser, E. T., Li, H. & Zhang, W. 2021 Bacterial community assembly and antibiotic resistance genes in the lettuce-soil system upon antibiotic exposure. *Science of the Total Environment* **778**, 146255.

Tian, P., Razavi, B. S., Zhang, X., Wang, Q. & Blagodatskaya, E. 2020 Microbial growth and enzyme kinetics in rhizosphere hotspots are modulated by soil organics and nutrient availability. *Soil Biology and Biochemistry* **141**, 107662.

Tu, Y., Li, H.-X., Jiang, L., Dong, K. & Wang, D.-Q. 2019 Bacterial communities structure and diversity in rhizosphere of different plants from Huixian wetland, Guangxi. *Ecology and Environmental Ences* **28** (2), 252–261.

Wang, M. & Helbling, D. 2016 A non-target approach to identify disinfection byproducts of structurally similar sulfonamide antibiotics. *Water Research* **102**, 241–251.

Wang, J., Chen, G., Fu, Z., Qiao, H. & Liu, F. 2021a Assessing wetland nitrogen removal and reed (*Phragmites australis*) nutrient responses for the selection of optimal harvest time. *Journal of Environmental Management* **280**, 111783.

Wang, J., Chen, X., Yan, W., Ning, C. & Gsell, T. 2021b Both artificial root exudates and natural *Koelreuteria paniculata* exudates modify bacterial community structure and enhance phenanthrene biodegradation in contaminated soils. *Chemosphere* **263**, 128041.

Wu, F.-Y., Chung, A. K. C., Tam, N. F. Y. & Wang, M. H. 2012 Root exudates of wetland plants influenced by nutrient status and types of plant cultivation. *International Journal of Phytoremediation* **14** (6), 543–553.

Wu, Y., Feng, P., Li, R., Chen, X., Li, X., Sumpradit, T. & Liu, P. 2019 Progress in microbial remediation of antibiotic-residue contaminated environment. *Chinese Journal of Biotechnology* **35** (11), 2133–2150.

Xia, F., Ma, D.-D., Zhang, J. & Wang, D.-Q. 2021 Characteristics and risk assessment of typical antibiotic contamination in Huixian wetland rivers. *Journal of Guilin University of Technology* **41** (1), 174–182.

Yan, Z., Hu, Y. & Jiang, H.-L. 2011 Toxicity of phenanthrene in freshwater sediments to the rooted submersed macrophyte, *Vallisneria spiralis*. *Bulletin of Environmental Contamination and Toxicology* **87**, 129–133.

Yan, Y., Chen, Y., Xu, X., Zhang, L. & Wang, G. 2019 Effects and removal of the antibiotic sulfadiazine by *Eichhornia crassipes*: potential use for phytoremediation. *Bulletin of Environmental Contamination and Toxicology* **103**, 342–347.

Yang, C. W., Hsiao, W. C. & Chang, B. V. 2016 Biodegradation of sulfonamide antibiotics in sludge. *Chemosphere* **150**, 559–565.

Yang, L., Barnard, R., Kuzyakov, Y. & Tian, J. 2021 Bacterial communities drive the resistance of soil multifunctionality to land-use change in karst soils. *European Journal of Soil Biology* **104**, 103313.

Yuan, Y., Zhao, W., Xiao, J., Zhang, Z., Qiao, M., Liu, Q. & Yin, H. 2017 Exudate components exert different influences on microbially mediated C losses in simulated rhizosphere soils of a spruce plantation. *Plant and Soil* **419**, 127–140.

Zhalnina, K., Louie, K. B., Hao, Z., Mansoori, N., da Rocha, U. N., Shi, S., Cho, H., Karaoz, U., Loqué, D., Bowen, B. P., Firestone, M. K., Northen, T. R. & Brodie, E. L. 2018 Dynamic root exudate chemistry and microbial substrate preferences drive patterns in rhizosphere microbial community assembly. *Nature Microbiology* **3** (4), 470–480.

Zhang, Y., Lin, S. S., Dai, C., Shi, L. & Zhou, X. F. 2014 Sorption desorption and transport of trimethoprim and sulfonamide antibiotics in agricultural soil: effect of soil type, dissolved organic matter, and pH. *Environmental Science and Pollution Research* **21**, 5827–5835.

Zhang, R., Pei, J., Zhang, R., Wang, S., Zeng, W., Huang, D., Wang, Y., Zhang, Y., Wang, Y. & Yu, K. 2018 Occurrence and distribution of antibiotics in mariculture farms, estuaries and the coast of the Beibu Gulf, China: bioconcentration and diet safety of seafood. *Ecotoxicology and Environmental Safety* **154**, 27–35.

Zhi, D., Yang, D., Zheng, Y., Yang, Y., He, Y., Luo, L. & Zhou, Y. 2019 Current progress in the adsorption, transport and biodegradation of antibiotics in soil. *Journal of Environmental Management* **251**, 109598.

First received 5 October 2022; accepted in revised form 7 January 2023. Available online 21 January 2023

doi: 10.2166/wrd.2022.065

Fluorescence analysis for water characterization: measurement processes, influencing factors, and data analysis

Zi-Bo Jing, Wen-Long Wang*, Yu-Jia Nong, Ping Zhu, Yao Lu and Qian-Yuan Wu

State Environmental Protection Key Laboratory of Microorganism Application and Risk Control (SMARC), Guangdong Provincial Engineering Research Center for Urban Water Recycling and Environmental Safety, Institute of Environment and Ecology, Tsinghua Shenzhen International Graduate School, Tsinghua University, Shenzhen 518055, China
*Corresponding author. E-mail: wwl20@sz.tsinghua.edu.cn

ABSTRACT

Fluorescence analysis is a sensitive and selective method that provides abundant information and does not result in sample destruction. This technology is widely used in the detection of dissolved organic matter in the environment. Some challenges with fluorescence analysis are its higher sensitivity so that it is sensitive to background signals, the difficulty of extracting useful information, and the complexity and diversity of analytical methods. This review summarizes recent applications of fluorescence analysis in water research for the characterization of pollutants, evaluation of water treatment processes, and monitoring of emerging contaminants such as drugs, disinfection by-products, and toxicity. Two-dimensional fluorescence and excitation–emission matrix fluorescence analysis methods are discussed, along with their advantages and disadvantages, and application scope. Methods for sample processing, instrument calibration, and data analysis are proposed. This review is an important source of information for the application of fluorescence technology in water research such as the analysis of emerging contaminants.

Key words: analysis, DOM, fluorescence, measurement

HIGHLIGHTS

- The application of fluorescence technology in water environments is summarized.
- The analytical methods and application scopes of two-dimensional fluorescence and EEM are discussed.
- A more accurate fluorescence analysis flow to control measurement and analysis errors is proposed.

GRAPHICAL ABSTRACT

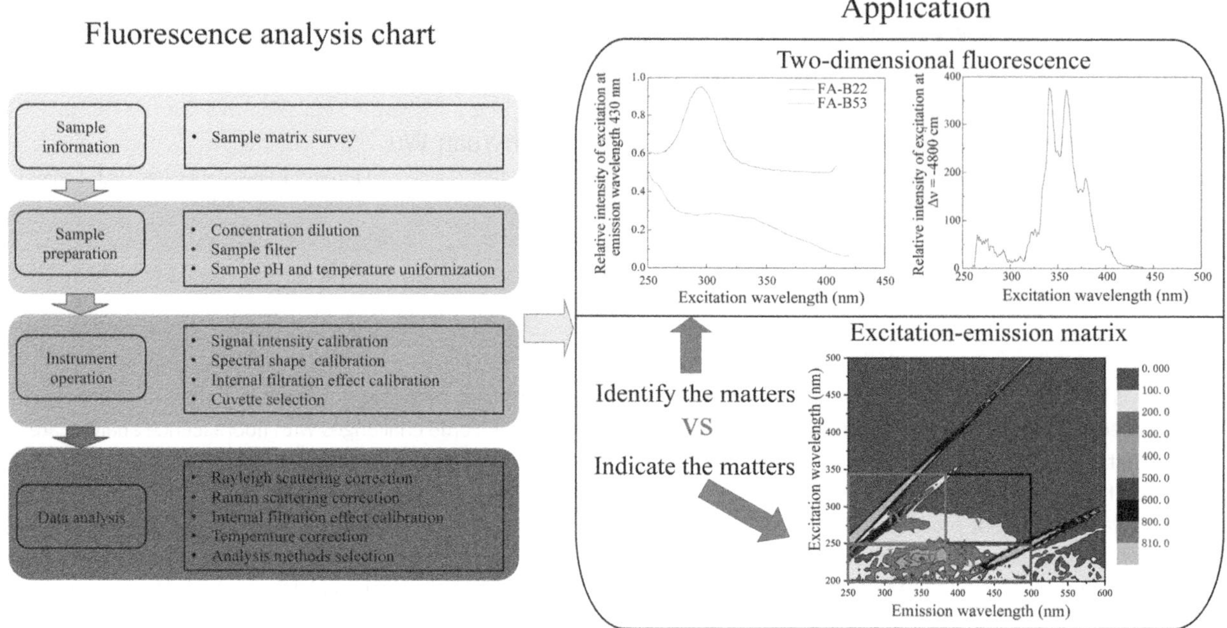

1. INTRODUCTION

Dissolved organic matter (DOM) is a complex mixture of humic acid, carbohydrates, amino acids, proteins, other natural organic matter (NOM), and synthetic organic compounds, and is widely found in water systems (Berg *et al.* 2019; Shi *et al.* 2021). In wastewater treatment systems, DOM causes membrane fouling and the disinfection by-products (DBPs) are precursors that affect the process operation and have potential health risks, respectively (Yang *et al.* 2015b; Li *et al.* 2020; Bai *et al.* 2021; Liu *et al.* 2021; Ren *et al.* 2021). Water quality indexes such as the chemical oxygen demand (COD) and biochemical oxygen demand (BOD) are used to assess water quality (Sgroi *et al.* 2017; Li *et al.* 2020; Xu *et al.* 2020; Shi *et al.* 2021). However, these analyses require the pretreatment of samples and have long detection times, which means they are not suitable for the rapid characterization of organic compounds in water.

Rapid and convenient analytical methods to measure DOM in aquatic environments are required. Especially in recent years, emerging contaminants put forward higher requirements for the detection of organic pollutants in water. Optical spectroscopy methods, including ultraviolet–visible absorption spectrometry (Zhang *et al.* 2020b), infrared spectroscopy (Yang *et al.* 2015b), and fluorescence spectroscopy (Yang *et al.* 2015a), have been used to analyze DOM. Fluorescence analysis has good selectivity and high sensitivity, provides abundant information and does not involve the destruction of the samples. Consequently, fluorescence analysis methods are frequently used to analyze DOM in water systems (Westerhoff *et al.* 2001; Li *et al.* 2014). These methods can be used to analyze the type and concentration of DOM in water (Lee *et al.* 2015; Ji *et al.* 2018), and the conversion of DOM in the water treatment process (Rodriguez-Vidal *et al.* 2020). They can also be used to evaluate the water treatment efficiency, such as the removal rate of trace organic pollutants (Sgroi *et al.* 2017), and indicate toxicity changes and disinfection by-product variations in water (Shen *et al.* 2018; Chen *et al.* 2020; Wang *et al.* 2021; Xu *et al.* 2021; Huang *et al.* 2022).

Traditional two-dimensional (2D) fluorescence analysis techniques, such as fluorescence excitation spectroscopy, fluorescence emission spectroscopy, and synchronous fluorescence spectroscopy, are mostly used to identify and characterize one or more substances (Bruckman *et al.* 2012; Foudeil *et al.* 2015). Excitation–emission matrix (EEM) analysis methods, including the fluorescence index (FI), fluorescence regional integration (FRI), and parallel factor analysis (PARAFAC) (Sgroi *et al.* 2017; Li *et al.* 2020), can be used to trace the sources of pollutants, evaluate the effects of water treatment processes, and monitor pollutants. These methods have been applied to water quality analysis of rivers and lakes (Patel-Sorrentino *et al.* 2002; Zhang *et al.* 2020a), groundwater (Vera *et al.* 2017), drinking water (Xu *et al.* 2021), municipal wastewater (Li *et al.* 2014; Mao *et al.* 2021), reclaimed water, and industrial wastewater (Rodriguez-Vidal *et al.* 2020; Islam *et al.* 2021). However, there are still many uncertainties about the process of fluorescence data analysis. High sample concentrations can exceed the analytical range, and extraction of useful information from complex multi-dimensional fluorescence data is difficult. In addition, water quality parameters and interactions between different components in the samples can affect fluorescence analysis.

In this paper, we aimed to (1) introduce the mechanism of fluorescence methods and the development of its application in water systems, (2) review 2D fluorescence and excitation–emission fluorescence methods and the application scope, and (3) identify and analyze the errors in fluorescence analysis to establish a rigorous fluorescence analysis process.

2. FLUORESCENCE APPLICATION

Fluorescence is a photoluminescence phenomenon and its general principles can be illustrated using a Jablonski diagram (Figure 1) (Noomnarm & Clegg 2009; Khan *et al.* 2022). Molecules in the ground state absorb excitation light (high energy photons, xenon light), photon energy is transferred to the molecules, and they enter an excited state. The excited molecules return to the ground state by non-radiative transitions (emission of heat or kinetic energy) and photoluminescence (emission of fluorescence and phosphorescence). The ground state (S_0) absorbs lower wavelengths of light (λ_1) and enters into the second excited singlet state (S_2). Alternatively, S_0 absorbs higher wavelengths of light (λ_2) and enters into the first excited singlet state (S_1). The higher energy second excited singlet state (S_2) can enter into the lower energy first excited singlet state (S_1) by internal conversion, and the first excited singlet state (S_1) can then return to the ground state by emission of fluorescence (λ_3). The first excited singlet state (S_1) may also enter the first excited triplet state (T_1) by intersystem crossing and then return to the ground state by emission of phosphorescence (λ_4). Vibrational relaxation occurs in the same electron energy level and transitions from high vibrational energy to low vibrational energy using energy exchange. According to their energies, the wavelengths are in order $\lambda_1 > \lambda_2 > \lambda_3 > \lambda_4$.

A search on the Web of Science for articles published between 2012 and 2021 with 'fluorescence' as the topic and 'environmental sciences ecology, water resources, and geochemistry geophysics' as the research interests gave 13,782 hits. A large number of articles illustrates how fluorescence has attracted much interest over the past 10 years, especially the past 4 years (Figure 2). The co-occurrence of certain keywords with 'fluorescence' from 2012 to 2021 was explored by bibliometric analysis (Figure 3) (van Eck & Waltman 2010). Fluorescence, DOM, water quality, drinking water, wastewater, and PAR-AFAC are frequently encountered as keywords with fluorescence, and this shows that fluorescence technology is often used for DOM analysis in aqueous environments. The colors (purple, green, and yellow) of the dots in Figure 3 represent

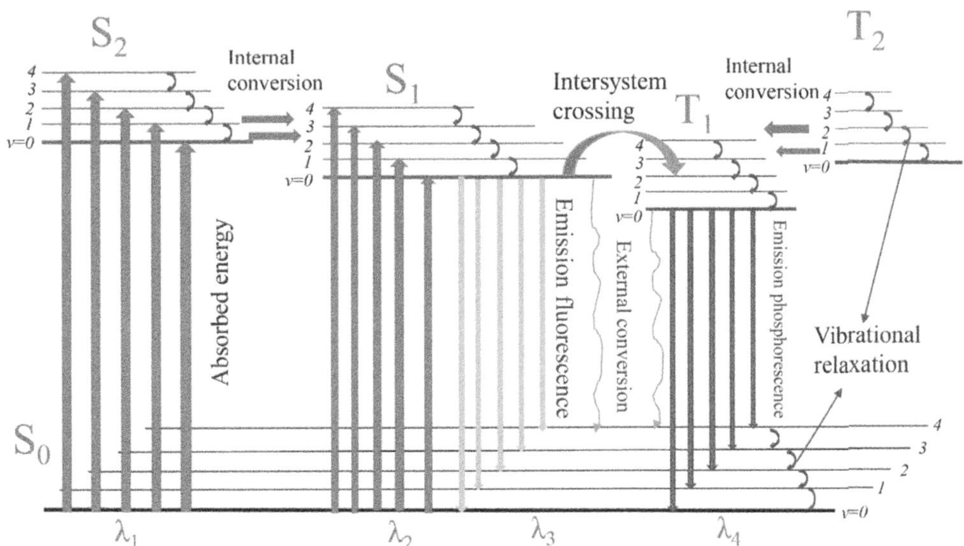

Figure 1 | Jablonski diagram showing energy transduction of different energy levels and fluorescence emission. S_0: ground state, S_1: first excited singlet state, S_2: second excited singlet state, T_1: first excited triplet state, T_2: second excited triplet state, λ_1: lower wavelength light, λ_2: higher wavelength light, λ_3: emission fluorescence, λ_4: emission phosphorescence (Noomnarm & Clegg 2009).

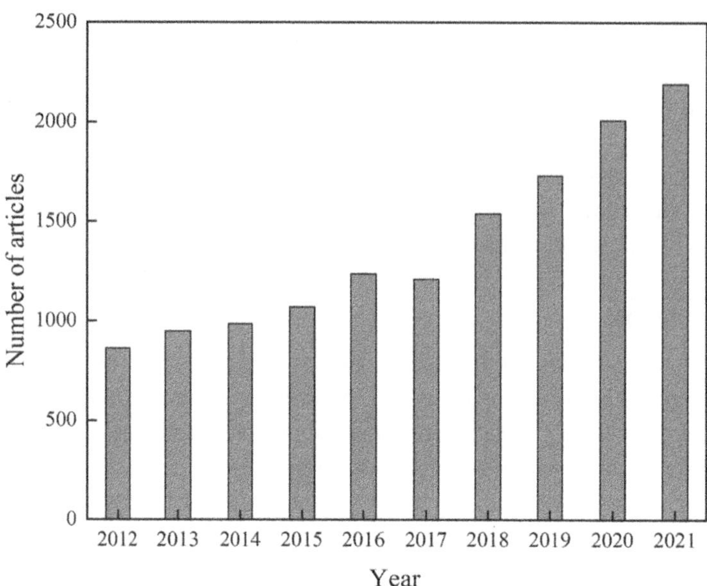

Figure 2 | Web of Science search of the number of articles about the topic of 'fluorescence' published between 2012 and 2021.

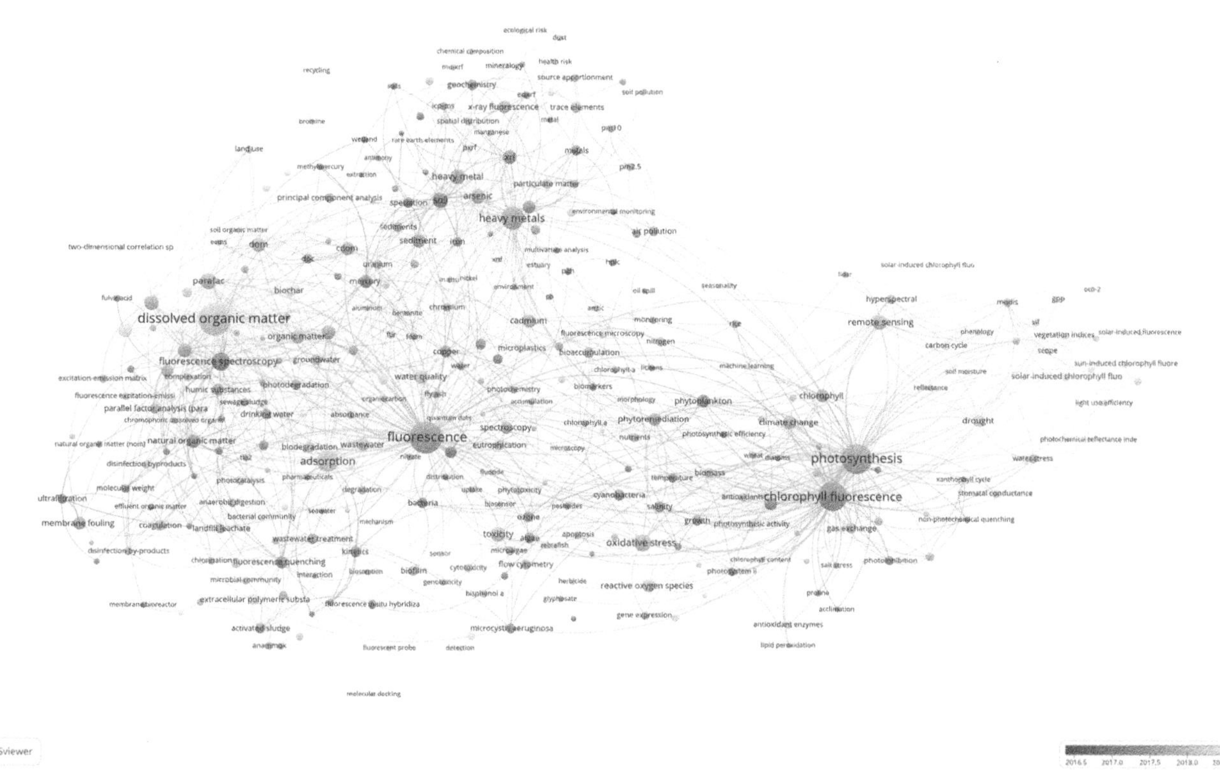

Figure 3 | Co-occurrence analysis of keywords related to fluorescence.

the order in which the keywords appear. DOM received more attention in recent years, and groundwater, drinking water, wastewater, EEM, humic substances, extracellular polymeric substances, and PARAFAC are closely related to DOM. PAR-AFAC is an important three-dimensional (3D) fluorescence analysis method. Fluorescence is often used to characterize DOM properties such as pollutant traceability and for water quality analysis and algal release analysis (Hanh *et al.* 2009;

Sgroi *et al.* 2017). It is also used to evaluate efficiency in drinking water treatment and sewage treatment (Li *et al.* 2020; Islam *et al.* 2021; Shi *et al.* 2021; Chen *et al.* 2022). Additionally, fluorescence can be used to indicate the emerging contaminants in water, such as trace organic contaminants, micropollutants, and DBPs (Yang *et al.* 2015b; Sgroi *et al.* 2017; Yadav *et al.* 2019; Fan *et al.* 2020).

3. 2D FLUORESCENCE SPECTRA

A 2D fluorescence spectrum represents the relationship between the fluorescence intensity and excitation wavelength or emission wavelength. 2D spectra are simple and intuitive with the *x*-axis representing the excitation wavelength or emission wavelength, and the *y*-axis representing the relative fluorescence intensity (Figure 4). From a 2D fluorescence spectrum, we can intuitively obtain the number and position of fluorescence peaks. These peaks can then be used to identify fluorescent substances or as the basis for selecting appropriate excitation and emission wavelengths for the fluorescence analysis of substances.

3.1. Fluorescence excitation and emission spectra

A fluorescence excitation spectrum is a plot of the excitation wavelength versus the fluorescence intensity obtained with a fixed emission wavelength (Figure 4(a)). Fluorescence excitation spectra show the fluorescence quantum yield at different excitation wavelengths. The peak position can be selected according to the sample's excitation spectrum when determining the sample's concentration or composition (Wakebe & Van Keuren 1999; Ma *et al.* 2011; Bruckman *et al.* 2012). The fluorescence excitation spectrum can be used to analyze the fluorescence characteristics of specific substances such as xanthene dyes (Wakebe & Van Keuren 1999), erythrosine (Ma *et al.* 2011), NOM (Abbt-Braun & Frimmel 1999), and fulvic acid from groundwater (Kumke *et al.* 1999). The process of absorption of energy by fluorescent substances is an excitation process, which means that the excitation and absorption spectra will have similar shapes, but the fluorescence spectrum is not completely equivalent to the absorption spectrum.

For the fluorescence emission spectrum, the fluorescence intensity is scanned under different emission wavelengths at a fixed excitation wavelength to obtain the relationship between fluorescence intensity and the emission wavelength (Figure 4(b)). Substances can be identified using fluorescence emission spectra. The emission spectrum of a fluorescent substance will show a Stokes shift (Reynolds 2003), where the emission and absorption spectra have the same shape but the emission wavelength will be greater than the excitation wavelength. This occurs because the molecule loses some energy through internal conversion and vibration relaxation when it reaches the first excited singlet state (or second excited singlet state) and returns to the ground state. The spacing of vibrational energy levels in the ground state is related to the shape of the emission spectrum. The ground state molecules will be in different vibrational energy levels after excitation, which causes the

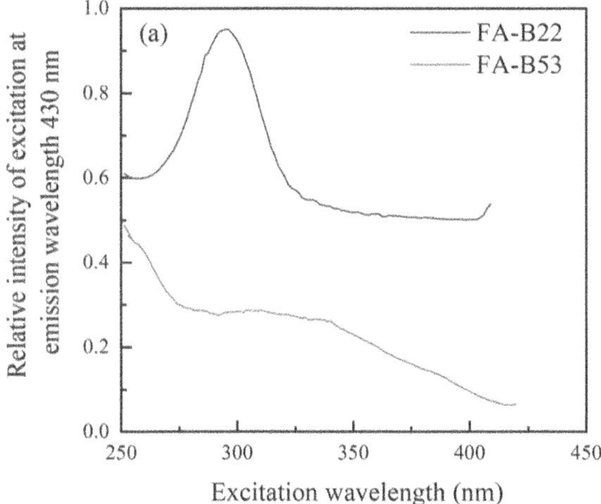

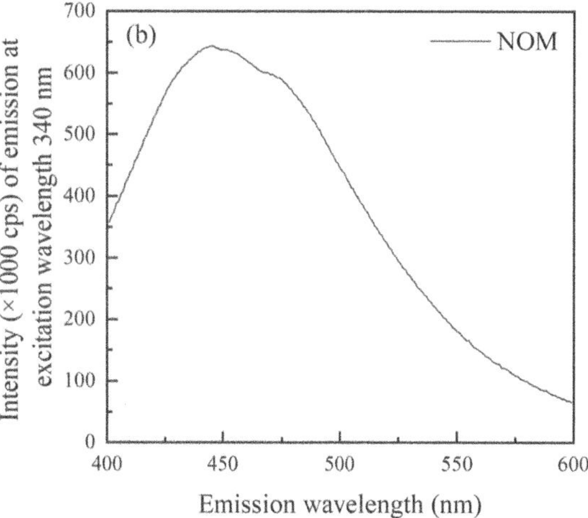

Figure 4 | Fluorescence excitation and emission spectra of (a) B22 fulvic acid and B53 fulvic at an emission wavelength of 430 nm with normalization of the spectra at an emission wavelength of 295 nm (Kumke *et al.* 1999), and (b) aquatic natural organic matter (NOM) at a fixed excitation wavelength of 340 nm (Chen *et al.* 2003a).

first absorption state in the absorption spectrum, and the ground state is always at the lowest vibrational energy level. Therefore, the shape of the first absorption band is related to the distribution of vibrational energy levels after excitation. According to the Franck–Condon principle (Stuhlmann *et al.* 2014), the probability of reaching a certain vibrational level is high during the excitation transition, and the probability of reaching the same vibrational level is also high during the radiative transition. Therefore, the fluorescence emission spectrum and absorption spectrum are mirror images. The independence of the emission spectrum from the excitation wavelength is mainly determined by the energy difference between the lowest vibrational energy level of the first excited singlet state and each vibrational energy level of the ground state.

Fluorescence emission spectra have been used to identify substances (Li *et al.* 1996; Zsolnay *et al.* 1999), analyze the fluorescence of NOM (Chen *et al.* 2003a) and DOM (Lombardi & Jardim 1999; Wei *et al.* 2005), and study the interactions of other factors with organic matter in the environment (Lu & Jaffe 2001; Gadad & Nanny 2008).

3.2. Synchronous fluorescence spectrum

For complex mixtures, such as sewage containing a variety of pollutants, the spectrum obtained by conventional fluorescence spectroscopy (emission and excitation spectroscopy) will be a combination of individual molecular peaks, and this will result in featureless spectral bands (Samokhvalov 2020). Synchronous fluorescence spectroscopy (SFS) is based on synchronous scanning technology proposed by Lloyd (1971). This method has been used in the identification and quantitation of complex mixtures (Vo-Dinh 1978). In SFS, the excitation and emission wavelengths are scanned at the same time, and a spectrogram is constructed from the measured fluorescence intensity and excitation wavelength (emission wavelength) (Figure 5).

The main types of SFS are constant-wavelength SFS (CWSFS), constant-energy SFS (CESFS), variable-angle SFS, and matrix isopotential SFS. In CWSFS, the excitation wavelength and emission wavelength are kept at a constant interval

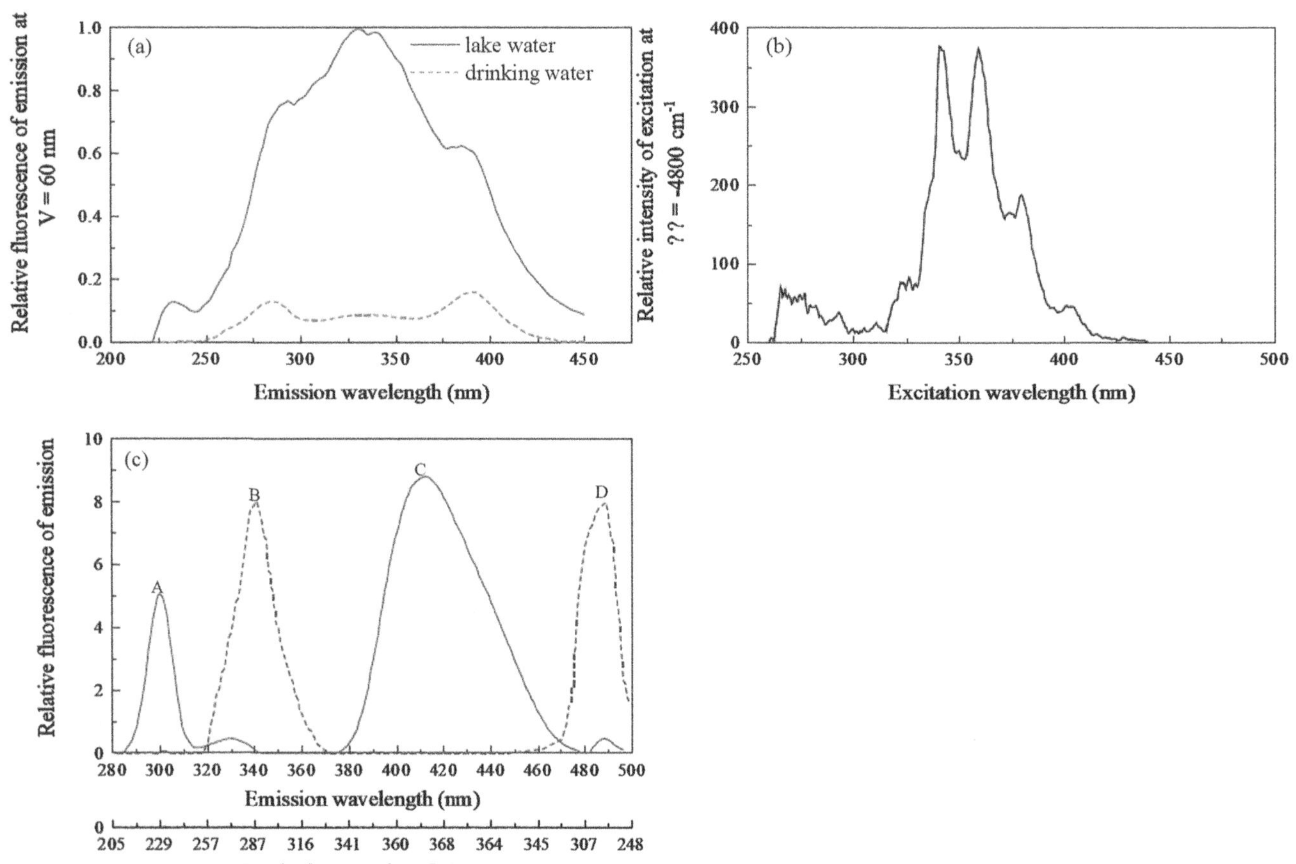

Figure 5 | (a) Constant-wavelength synchronous fluorescence spectra ($\Delta\lambda = 60$ nm) of lake water and drinking water (Reynolds 2003). (b) Constant-energy synchronous fluorescence spectra ($\Delta v = -4{,}800$ cm^{-1}) of a mixture of 16 PAHs (Yang *et al.* 2008). (c) Non-linear variable-angle synchronous fluorescence spectra for atenolol (A), propranolol (B), amiloride (C) and dipyridamole (D).

($\Delta\lambda = \lambda_{em} - \lambda_{ex}$) during fluorescence scanning (Figure 5(a)) (Lloyd 1971). CESFS was proposed by Inman & Winefordner (1982) as a method to keep a constant excitation energy and emission monochromator during wavelength scanning (Figure 5(b)). CESFS is based on the specific vibration energy of the molecule. If the selected constant energy difference is equal to the vibrational energy difference, and the excitation energy and emission energy are within the vibrational energy difference, a synchronous spectrum with the maximum intensity is generated. Compared with CWSFS, CESFS can effectively overcome the Raman scattering and improve the analytical sensitivity (Andrade-Eiroa *et al.* 2010). Compared with variable angle SFS, constant energy, and constant wavelength scans are performed with constant separation between emission and excitation beams (Figure 5(c)). The spectrum will be a straight line with a slope of one and the axes in nanometers or reciprocal centimeters. The scan path of variable angle SFS is a straight line with varying slopes (Andrade-Eiroa *et al.* 2010). The nonlinear scan path is a broken line or curve in the EEM diagram. Cabaniss (1991) distinguished variable angle SFS by the slope and intercept, which are defined by wavelength and energy. Matrix isopotential SFS was proposed by Pulgarin & Molina (1994) as a nonlinear variable angle SFS method. In this method, the fluorescence intensity line of the matrix can be used to eliminate the influence of the matrix.

SFS is a well-known and well-established technology that has been widely used for the analysis of water and wastewater samples (Pullin & Cabaniss 1995; Reynolds 2003; Andrade-Eiroa *et al.* 2010). It has been applied to the detection of organic substances such as polycyclic aromatic hydrocarbons (PAHs) and pesticides in water (Rodriguez & Sanz 2000; Foudeil *et al.* 2015), the determination of humic and fulvic acids (Peuravuori *et al.* 2002), wastewater fingerprinting (Wu *et al.* 2006), and analysis of water toxicity (Hanh *et al.* 2009). Although SFS has been widely used in various fields, its application is limited by the internal filtration effect, fluorescence quenching, Raman and Rayleigh scattering, and the strong superposition effect of fluorescence signals from multi-contaminant matrix samples.

4. EEM

Compared with 2D fluorescence, the main difference with EEM is that changes in the fluorescence intensity information for the excitation and emission can be obtained simultaneously. Contour fluorescence spectrograms or 3D projections are commonly used to display EEM data. Coble *et al.* (1990) reported the first application of an EEM to analyze the fluorescence characteristics of DOM in the Black Sea. Currently, EEM is the main method used to characterize the source, composition, and other information of DOM. However, EEMs containing multi-dimensional information are often difficult to analyze to achieve the research objectives (Sgroi *et al.* 2017; Li *et al.* 2020). The three simplified methods for EEM data analysis are the FI, FRI, and PARAFAC.

4.1. FI

The FI is the ratio of fluorescence intensities at emission wavelengths of 470 and 520 nm (FI = F_{470}/F_{520}) when the excitation wavelength is 370 nm. This method can be used to distinguish the source of DOM because an FI > 1.9 indicates strong microbial action, an FI < 1.4 indicates the DOM is from terrestrial or soil sources, and an FI of 1.4–1.9 indicates that the DOM is from a combination of terrestrial and autogenous sources (McKnight *et al.* 2001).

The biological index (BIX) and humification index (HIX) have also been developed as indicators of DOM (Rodriguez-Vidal *et al.* 2020). The BIX is calculated as the ratio of fluorescence intensities with emission wavelengths of 380 and 430 nm (BIX = F_{380}/F_{430}) at an excitation wavelength of 245 nm. A BIX > 1 indicates that biological sources are mainly influenced by organisms, and a BIX of 0.6–0.7 indicates that terrestrial input or human activities greatly affect biological sources (Huguet *et al.* 2009). The HIX is the ratio of the average light intensity in the range of 435–480 nm and 300–345 nm when the excitation wavelength is 245 nm. The humification degree increases with the HIX (Morling *et al.* 2017).

The FI method has been widely used to characterize organic matter sources in water research (Table 1). The FI, BIX, and HIX have been used to distinguish the sources of DOM (microbial or terrestrial) (Rodriguez-Vidal *et al.* 2020), colored DOM in wastewater (Clark *et al.* 2020), and DOM in lakes (Carstea *et al.* 2014; Zhang *et al.* 2020b). The FI has also been used to evaluate the water quality indicators of COD, 5-day BOD, total nitrogen, and ammoniacal nitrogen (Zhang *et al.* 2020a). Moreover, it has been shown that the FI can be used to evaluate emerging contaminants, such as DBPs, and their potential for formation (Yang *et al.* 2015b; Shen *et al.* 2018; Xu *et al.* 2021).

4.2. FRI

An EEM contains tens of thousands of excitation–emission wavelength-dependent fluorescence intensity dots. The volume of data makes analysis difficult. To address this, Chen *et al.* (2003b) proposed dividing the EEM into five EEM regions using a

Table 1 | Summary of the application of EEM fluorescence analysis methods

Analysis methods	Water type and treatment process	Excitation/Emission	Indication	Reference
FI	Drinking water	BIX	Soluble microbial products and disinfection by-products formation potentials	Shen *et al.* (2018)
	Lake	FI	Origin of chromophoric dissolved organic matters as terrestrial humic-like substances	Carstea *et al.* (2014)
	Drinking water treatment plant	BIX, HIX	Total trichloromethane formation potentials	Yang *et al.* (2015b)
	Drinking water	FI, BIX and HIX	Dissolved organic matters content and its disinfection by-products formation potential	Xu *et al.* (2021)
	Lake	FI, BIX and HIX	The concentration of chemical oxygen demand, biochemical oxygen demand, total nitrogen and ammonia nitrogen	Zhang *et al.* (2020a)
	Wastewater	FI, BIX and HIX	Origin of dissolved organic matters	Rodriguez-Vidal *et al.* (2020)
	Wastewater treatment plant	FI, BIX	Origin of chromophoric dissolved organic matters	Clark *et al.* (2020)
	Rainwater	HIX	Humification degree	Yang *et al.* (2019)
	Lake	FI, HIX	Origin of dissolved organic matters	Zhang *et al.* (2020b)
FRI	Drinking water	Regions III, V	Disinfection by-products and disinfection by-products formation potentials	Trueman *et al.* (2016)
	Drinking water treatment plant	Regions III, VI, V	Disinfection by-products	Fan *et al.* (2020)
	Drinking water	Region IV	Soluble microbial products and disinfection by-products formation potentials	Shen *et al.* (2018)
	River	Regions II, IV	Nitrogenous disinfection byproduct	Tan *et al.* (2017)
	Drinking water treatment plant	Regions II, IV	Nitrogenous disinfection byproduct	Lin *et al.* (2019)
	Wastewater treatment plan	Regions IV, V	Disinfection by-products	Liu *et al.* (2016)
	Wastewater	Regions II, III	Toxicity evolution during UV-driven oxidation	Huang *et al.* (2022)
	Wastewater	Regions I, II, IV	Toxicity evolution during up-flow anaerobic sludge blanket	Chen *et al.* (2020)
	Wastewater	$250 \sim 300$ nm/ > 320 nm	The concentration of chemical oxygen demand	Wang *et al.* (2022)
	Lake	Regions III, VI, V	The concentration of chromophoric dissolved organic matters	Ji *et al.* (2018)
	Wastewater treatment plan	Regions III, V	The concentration of chromophoric dissolved organic matters	Islam *et al.* (2021)
	Wastewater treatment plan	Regions V	Evaluation of emerging trace organic compounds removal	Sgroi *et al.* (2017)
	Landfill leachate	Regions II, V	The contamination of groundwater by landfill leachate	He & Fan (2016)
	Wastewater treatment plan	Regions II	Evaluation of antibiotic removal	Yadav *et al.* (2019)
PARAFAC	Wastewater treatment plan	C2 275 /340 nm	Evaluation of antibiotic removal	Yadav *et al.* (2019)
	Wastewater treatment plan	245, 350/450 nm; < 240, 315/380	Evaluation of emerging trace organic compounds removal	Sgroi *et al.* (2017)
	Drinking water treatment plant	C1 245/414 nm C2 230 (280)/328 nm	Nitrosamine	Maqbool *et al.* (2020)
	Drinking water treatment plant	C2 355/454.5 nm C3 250, 280/354.5 nm	Trichloromethane and para nitroso dimethylaniline formation potentials	Yang *et al.* (2015b)
	Bench-scale experiment	C1 260/455 nm C2 220/435 nm C3 225, 270/390 nm	The removal of humic substances by coagulation/flocculation process	Aftab & Hur (2017)

(Continued.)

Table 1 | Continued

Analysis methods	Water type and treatment process	Excitation/Emission	Indication	Reference
	Drinking water	C1 250(310)/425 nm C2 260(380)/480 nm C4 260/440 nm	Disinfection by-products formation potentials	Xu *et al.* (2021)
	Drinking water	C1 224 (314)/398 nm C2 344/466 nm C3 289/344 nm C4 279/294 nm	The performance of coagulation-filtration process	Sanchez *et al.* (2013)
	Lake	C1 260/430 nm C2 280 (430)/555 nm C3 240 (390)/480 nm C4 250/515 nm	The concentration of chemical oxygen demand, biochemical oxygen demand, total nitrogen and ammonia nitrogen	Zhang *et al.* (2020a)
	Wastewater	C1 240/421 nm C2 305/415 nm	The specific fingerprint of the wastewater	Rodriguez-Vidal *et al.* (2020)
	Rainwater	C1 345/437 nm C2 300/408 nm C3 (250,330)/456 nm C4 275/311 nm	The concentration of chromophoric dissolved organic matters	Yang *et al.* (2019)

method named FRI. FRI uses vertical or horizontal lines to divide the EEM fluorescence contour spectra into the following regions: I, aromatic proteins class I (tyrosine-like substances); II, aromatic proteins class II (tryptophan-like substances); III, fulvic acid-like substances; IV, soluble microbial by-product-like substances; and V, humic acid-like substances. It is worth noting that the EEM regions determined by the FRI method are constant, and the material itself or external conditions only slightly affect the DOM position (Li *et al.* 2020). As a semi-quantitative method, the fluorescence flux of different regions can be calculated by the volume integral to obtaining two indexes. These indexes are the volume ($\Phi_{i, n}$ for each region, $\Phi_{T, n}$ for the whole region) and percent fluorescence response ($P_{i, n}$), and can be used to evaluate differences between samples or the change for a substance within a single sample (Li *et al.* 2020).

Because FRI analysis is simple and stable, it has been widely applied to investigate the composition and variation of DOM in water and wastewater (Table 1). FRI is often used to track and determine the fate of organic matter during water treatment (Sgroi *et al.* 2017). FRI has also been used to evaluate the performance of processes at treatment facilities, such as ultrafiltration, biological activated carbon, ultraviolet advanced oxidation, and membrane bioreactors (Vera *et al.* 2017; Shen *et al.* 2018; Lin *et al.* 2019; Huang *et al.* 2022). Similar to the FI method, FRI can be used to indicate emerging contaminants (Sgroi *et al.* 2017) such as pharmaceuticals and personal care products (Yadav *et al.* 2019), DBPs (Trueman *et al.* 2016; Shen *et al.* 2018; Fan *et al.* 2020), nitrogenous DBPs (Tan *et al.* 2017; Lin *et al.* 2019), and toxic compounds (Chen *et al.* 2020; Huang *et al.* 2022) in addition to detecting conventional indicators of water quality (Wang *et al.* 2022). Compared with the FI method, FRI is beneficial for application to substances with similar structures.

4.3. PARAFAC

For EEM with complex data structures, multi-way data analysis methods (chemometrics) are required. PARAFAC is the main chemometrics method used to process EEM data and has been applied to the analysis of DOM in water and wastewater (Sgroi *et al.* 2017; Li *et al.* 2020). PARAFAC is a quantitative and qualitative analysis method. It decomposes complex EEM fluorescence data into individual fluorescence phenomena and has become the most commonly used EEM analysis method. The three-way dataset is decomposed into three loading matrices by the PARAFAC model as shown in Equation (1) (Bro 1997; Stedmon & Bro 2008).

$$x_{ijk} = \sum_{f=1}^{F} a_{if} b_{jf} c_{kf} + \varepsilon_{ijk}, \ i = 1\ldots I, j = 1\ldots J, k = 1\ldots K, \tag{1}$$

where x_{ijk} is one element of a three-way data array with dimensions I, J, and K. In the analysis of EEMs, x_{ijk} is the fluorescence intensity of sample i measured at emission wavelength j, and excitation wavelength k. The final term ε_{ijk} represents the unexplained signal (residuals containing noise and other unmodeled variations). The outcomes of the model are the parameters *a*, *b*, and *c*. Ideally, *a*, *b*, and *c* represent the concentration, emission spectra, and excitation spectra of the underlying fluorophores, respectively. The stages of PARAFAC analysis are data consideration, preliminary data treatment, exploration of the dataset, validation of the model, and interpretation of the results (Stedmon & Bro 2008). For the data consideration stage, approximately 20–100 samples are required for EEM PARAFAC analysis and established models cannot be extrapolated to other data for analysis. Instrument calibration and removal of Raman and Rayleigh scattering are the main steps for preliminary data treatment. The drEEM toolbox in MATLAB can be used to decompose the dataset, validate the model, and interpret the results (Hambly *et al.* 2015).

Several representative components are usually obtained by EEM PARAFAC analysis of samples in aquatic environments (Table 1). These components can be used instead of conventional water quality indicators, such as the COD, BOD_5, ammoniacal nitrogen, and total phosphorus (Zhang *et al.* 2020a), and emerging contaminants, such as DBPs, nitrogenous DBPs, *N*-nitrosamines, and pharmaceuticals and personal care products (Yang *et al.* 2015b; Sgroi *et al.* 2017; Yadav *et al.* 2019; Maqbool *et al.* 2020; Xu *et al.* 2021). These components can also be used to assess the effectiveness of water treatment technologies and the fate of organic matter in water treatment processes (Sanchez *et al.* 2013; Aftab & Hur 2017; Sgroi *et al.* 2017; Rodriguez-Vidal *et al.* 2020).

5. FACTORS INFLUENCING THE FLUORESCENCE SPECTRUM MEASUREMENT AND ANALYSIS

The fluorescence produced by a substance is related to the substance's structure; however, equipment, operational parameters, and environmental factors can affect the shape and size of the fluorescence spectrum. Therefore, the sensitivity and selectivity of fluorescence analysis can be improved by selecting appropriate operating conditions.

5.1. Influence of operational parameters

In all fluorescence measurement methods, instrument calibration is a necessary step to correct instrument biases. Instrument calibration includes the signal intensity and spectral shape. For example, for DOM measurement, the fluorescent molecules or fluorophores emitting fluorescence are unknown, so it is necessary to calibrate the signal intensity to compare different samples. The most common calibration method involves measuring quinine sulfate at an excitation wavelength of 350 nm and emission wavelength of 450 nm (Coble *et al.* 1993). The Raman signal of pure water can also be used for correction (Determann *et al.* 1994; Stedmon *et al.* 2003). Spectral correction takes into account the spectral output deviation of the instrument light source and biases of the instrument light transmission components. With the development of instrument technology, most currently available fluorescence measuring devices have a built-in reference detector to correct the optical signal for the light source spectrum. Residual deviations can be corrected periodically with rhodamine-B or rhodamine-101 (Karstens & Kobs 1980). Emission spectra also need to be corrected, but they generally show little variation and correction can generally be performed by changing the instrument operation manually.

Scattered light, such as that from Rayleigh and Raman scattering, has a large influence on fluorescence measurement. Molecular absorption of low photon energy can only excite electrons in the molecule to other higher vibrational levels of the ground state, but not to a higher excited state. In Rayleigh scattering, an excited electron that loses no energy quickly returns to its original ground state and emits light in which the radiation wavelength is the same as the excitation wavelength. In other vibrational levels of the ground state, the electrons in the molecule do not return to the initial ground state, but return to a higher or lower level than the original vibrational level. At this time, the wavelength of radiated light is longer or shorter than the excitation wavelength, and the radiated light is Raman scattered. Therefore, the occurrence of Rayleigh and Raman scattering in fluorescence measurements limits the sensitivity and reliability of fluorescence analysis. Emission gratings with a double monochromator and a cut-off filter in the measuring instrument can attenuate the effect of scattered light (Murphy *et al.* 2013). Raman scattering can be removed by deducting the pure water spectrum from the sample spectrum (Stedmon & Bro 2008). In the Rayleigh scatter-affected region, scattered signals are often treated as missing data to remove the effects of scattered light (Christensen *et al.* 2003; Stedmon & Bro 2008), and a blank fluorescence spectrum is used for subsequent fluorescence modeling and analysis. The above problems can be avoided by inserting zeros outside of the data region (Rinnan *et al.* 2005). The *smootheem* functions in the drEEM toolbox in MATLAB can be used to remove scattered signals or interpolate to remove Rayleigh and Raman scattering (Murphy *et al.* 2013).

5.2. Influence of water parameters

Because of the absorption of excited or emitted fluorescent photons by the sample matrix, the inner filter effect (IFE) reduces the fluorescence yield. The main reason for the IFE is that some chromophores in the sample absorb photons at the same wavelength as the target molecule (Chen *et al.* 2018). There are two types of IFE: primary IFE involves the absorption of excited photons by the fluorophores and chromophores; and secondary IFE involves the absorption of photons emitted by fluorophores and chromophores (Ohno 2002). IFE will distort the fluorescence spectrum and adversely affect the fluorescence analysis of substances. Generally, analysis of samples with low concentrations is considered to have a low IFE. However, the IFE should be corrected for accurate analysis of samples at high concentrations (Panigrahi & Mishra 2019). The IFE calibration methods include instrument calibration, parameter correction, and mathematical correction (Khan *et al.* 2022). Instrument calibration is accomplished by instrument theory, parameter correction uses absorbance, optical density, and other parameters for correction, and mathematical correction is used for subsequent fluorescence data analysis (Ohno 2002; Chen *et al.* 2018).

The emission fluorescence characteristics of some molecules with acidic or fluorophores can be considered as two types. A variation in pH will change the proportions of two fluorescent molecules, which will affect the position and shape of the fluorescence spectrum and the fluorescence intensity. For example, a study of the fluorescence spectra of humic substances from the International Humic Substances Society showed that the fluorescence spectra of samples redshifted with an increase in pH (Pullin & Cabaniss 1995). This shift was attributed to a change in the fluorescence characteristics of acidic functional groups in humic substances (Mobed *et al.* 1996). Moreover, it has been verified that the fluorescence intensity increases with increasing pH. This has been observed in investigations of organic matter in river water and the detection of extracellular organic matter under different pH conditions (Patel-Sorrentino *et al.* 2002; Sheng & Yu 2006). When the pH changes from neutral to alkaline, the fluorescence intensity of treated wastewater decreases by 30–40% (Westerhoff *et al.* 2001). Similarly, salinity can also change the position and shape of the fluorescence spectrum and the fluorescence intensity by changing the molecular characteristics of fluorescent molecules (Khan *et al.* 2022). For example, one study found that the fluorescence intensity increased by approximately 10% when the pH was increased from 7.0 to 8.5 and the salinity was doubled (Esteves *et al.* 1999).

Increases in the temperature have an adverse effect on fluorescence because the probability of collisions between molecules increases, which increases the probability of de-excitation (Carstea *et al.* 2014; Lee *et al.* 2015). The fluorescence intensities of humic substances from the International Humic Substances Society, tryptophan standard, river water, and sewage were quenched when the temperature was increased from 10 °C to 45 °C (Baker 2005). Elliott *et al.* (2006) observed that the fluorescence decreased by more than 40% as the temperature increased. The degree of thermal quenching depends on the type of water because the fluorophores in water come from different sources. Terrestrial humic-like components in rural water samples cause higher quenching than those in urban water samples (Carstea *et al.* 2014). Thus, it is necessary to correct the influence of temperature on the fluorescence spectrum in DOM fluorescence analysis. Temperature correction tools based on sequential mathematical correction methodology have been used to address the effect of temperature on fluorescence measurements (Watras *et al.* 2011; Goffin *et al.* 2020).

Metal ions can chelate with DOM or precipitate DOM to enhance or quench fluorescence. There are two types of DOM fluorescence quenching by metal ions. The first, which is called dynamic quenching, involves the formation of stable non-fluorescent complexes between metal ions and DOM fluorescence binding sites. The second, static quenching, involves the formation of metal complexes that have only partial fluorescence because of the inherent chemical heterogeneity of DOM (i.e., different chemical structures and/or similar chemical structures in different environments) (da Silva *et al.* 1998; Yamashita & Jaffe 2008). Static quenching decreases with temperature increases, whereas dynamic quenching increases with increases in temperature (Khan *et al.* 2022). In dynamic quenching, non-fluorescent metal complexes deactivate excited molecules through intermolecular or intramolecular collisions (Lakowicz 2006). As the temperature increases, it is difficult for the molecules in the ground state to enter the excited state, which weakens the fluorescence signal. Paramagnetic metal ions (Cu^{2+}, Fe^{3+}, Hg^{2+}, Ni^{2+}, and Zn^{2+}) can quench the fluorescence of humic-like substances but not protein-like substances (Provenzano *et al.* 2004; Henderson *et al.* 2009). Diamagnetic metal ions (Al^{3+}, Mg^{2+}, Ca^{2+}, and Cd^{2+}) show different effects on fluorescence (i.e., enhancement, quenching, or minimal effect) (Elkins & Nelson 2002). Tryptophan-like fluorescence is quenched by Cu^{2+}, Ni^{2+}, Fe^{3+}, and Mo^{3+}, but not by Mn^{2+}, Co^{2+}, Ca^{2+}, Zn^{3+}, Cr^{3+}, and Na^+ (Henderson *et al.* 2009). Different concentrations of metal ions and fluorophores, pH values, and temperatures in various systems lead to different levels of fluorescence quenching (Henderson *et al.* 2009; Pan *et al.* 2012). Therefore, for fluorescence measurements in complex matrices such as sewage, the influence of different components should be considered.

The solvent effect is another important factor affecting fluorescence analysis (Mataga *et al.* 1955). Under the influence of the dielectric constant and refractive index of a solution, the fluorescence intensity and maximum wavelength of the fluorescence spectrum are generally affected by the solvent. Hydrogen bonding between the solvent and fluorescent molecules can also cause the solvent effect (Siqintuya *et al.* 2005). The polarity of the solvent is the main contributor to the solvent effect on the fluorescence properties of a substance (Xie *et al.* 2004; Qiu *et al.* 2010). The complexation of ground and excited

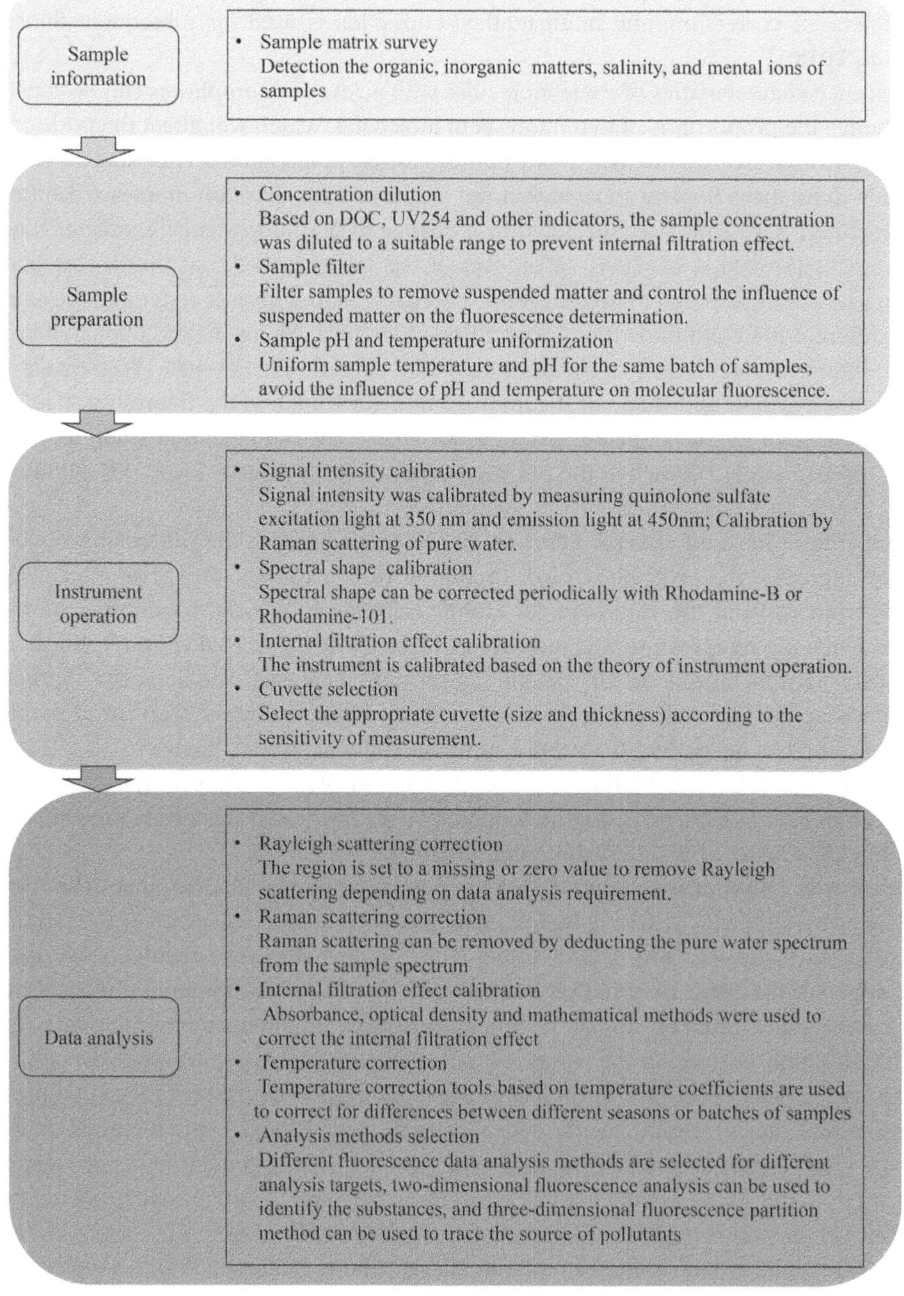

Figure 6 | Flow chart of fluorescence measurement data analysis. Derived from the literature: (Karstens & Kobs 1980; Coble *et al.* 1993; Pullin & Cabaniss 1995; Ohno 2002; Christensen *et al.* 2003; Stedmon *et al.* 2003; Baker 2005; Stedmon & Bro 2008; Watras *et al.* 2011; Chen *et al.* 2018; Panigrahi & Mishra 2019; Khan *et al.* 2022).

molecules and polar molecules enhances fluorescence (Rechthaler & Kohler 1994). The fluorescence spectra of quinolone antibiotics in different solvents showed that the dielectric effect of the solvent was stronger than that of specific hydrogen bonding (Park *et al.* 2002). Therefore, the solvent effect should be considered when fluorescence technology is used to identify certain substances or analyze the removal of characteristic pollutants.

5.3. Flow diagram of fluorescence measurement and data analysis

The accuracy, stability, and sensitivity of fluorescence measurement and data analysis can be affected by various factors. Consequently, calibration is required during experimental and analytical processes. For the measurement and analysis of EEM fluorescence data as an example, we developed a flow diagram to represent considerations in fluorescence measurement and analysis (Figure 6). First, the organic and inorganic components, salinity, and ionic strength of the samples should be investigated and analyzed, along with their effects on the target substances. Second, for sample preparation before measurement, the pH and temperature should be adjusted. In addition, the sample concentration should be diluted to a suitable level to avoid IFE. Especially, solvent effects should be considered when using 2D spectroscopy for the identification of substances or in laboratory-scale removal studies for specific contaminants. Third, instrument calibration should be performed to calibrate the signal intensity, spectral shape, and IFE during measurement. Finally, for analysis of the data after fluorescence measurement, the IFE should be removed by mathematical analysis, the temperature correction coefficient should be used to correct the temperature effect (e.g., for comparison of samples from different seasons), and Raman and Rayleigh scattering should be corrected. There are many other environmental factors or operating conditions that could affect the fluorescence spectrum, and these still need to be elucidated and solved.

6. CONCLUSIONS

Fluorescence analysis has been widely used to characterize DOM in water and wastewater and used to trace pollutants, assess the performance of water treatment technologies, and indicator characteristic pollutants including emerging contaminants (e.g., drugs and DBPs).

There are two main categories of fluorescence analysis methods: 2D methods, such as fluorescence excitation and emission spectroscopy and synchronous fluorescence spectroscopy; and EEM methods of the FI, FRI, and PARAFAC. The 2D analysis methods are mainly limited to the identification of one or several pollutants. By contrast, EEM analysis methods provide more complex fluorescence information and can be used to analyze a variety of pollutants. Consequently, EEM analysis has become the most widely used technology.

Moreover, this review evaluated a variety of methods to reduce the error in fluorescence analysis and developed a flow chart for fluorescence analysis. The error in fluorescence analysis is affected by two types of parameters: operation parameters, including the instrument calibration, cuvette selection, and removal of Raman and Rayleigh scattering; and water quality parameters, including the IFE, pH, temperature, metal ions, and solvent effect. This analysis flow chart will support future research on improving the accuracy of fluorescence analysis.

7. RECOMMENDATIONS FOR FURTHER STUDIES

Although fluorescence technology has been widely used in water environments, the following issues are required to be further studied and analyzed in the future:

1. There is a need to conduct a series of studies to explore the relationship of emerging contaminants with fluorescence. With the continuous discovery of emerging pollutants, there is an increasing emphasis on their rapid characterization. Therefore, the exploration of fluorescence characteristics of pollutants is conducive to the rapid and convenient analysis of water quality.
2. Further studies are needed to analyze the influence limits of temperature, pH, metal ions, internal filtration effect, and other environmental factors, and to develop a more complete and standard fluorescence analysis process to obtain accurate fluorescence data.
3. Online water quality detection has become an important form of water quality analysis and display for drinking water treatment and supply, and wastewater treatment and discharge. Fluorescence has great advantages as a fast and sensitive method for the detection of organic matter. Furthermore, the correlation of fluorescence signals and conventional indicators such as COD and BOD, and emerging contaminants such as antibiotics, endocrine disruptors, and DBPs.

ACKNOWLEDGEMENTS

This study was supported by the National Natural Science Foundation of China (No. 52000115), Tsinghua Shenzhen International Graduate School (HW2021012/QD2021010C), Shenzhen Science, Technology and Innovation Commission (No. JCYJ20200109142829123), and Guangdong Basic and Applied Basic Research Foundation (2020A1515110106).

DATA AVAILABILITY STATEMENT

All relevant data are included in the paper or its Supplementary Information.

CONFLICT OF INTEREST

The authors declare there is no conflict.

REFERENCES

Abbt-Braun, G. & Frimmel, F. H. 1999 Basic characterization of Norwegian NOM samples – similarities and differences. *Environment International* **25** (2–3), 161–180.

Aftab, B. & Hur, J. 2017 Fast tracking the molecular weight changes of humic substances in coagulation/flocculation processes via fluorescence EEM-PARAFAC. *Chemosphere* **178**, 317–324.

Andrade-Eiroa, A., de-Armas, G., Estela, J.-M. & Cerda, V. 2010 Critical approach to synchronous spectrofluorimetry. I. *Trac-Trends in Analytical Chemistry* **29** (8), 885–901.

Bai, Y., Wu, Y.-H., Tong, X., Wang, Y.-H., Ikuno, N., Wang, W., Shi, Y.-L. & Hu, H.-Y. 2021 Revealing the membrane fouling mechanism caused by the denitrification filter effluent during ozonation by model assessment. *Water Reuse* **11** (2), 149–159.

Baker, A. 2005 Thermal fluorescence quenching properties of dissolved organic matter. *Water Research* **39** (18), 4405–4412.

Berg, S. M., Whiting, Q. T., Herrli, J. A., Winkels, R., Wammer, K. H. & Remucal, C. K. 2019 The role of dissolved organic matter composition in determining photochemical reactivity at the molecular level. *Environmental Science & Technology* **53** (20), 11725–11734.

Bro, R. 1997 PARAFAC. tutorial and applications. *Chemometrics and Intelligent Laboratory Systems* **38** (2), 149–171.

Bruckman, L. S., Richardson, T. L., Swanstrom, J. A., Donaldson, K. A., Allora, M., Shaw, T. J. & Myrick, M. L. 2012 Linear discriminant analysis of single-cell fluorescence excitation spectra of five phytoplankton species. *Applied Spectroscopy* **66** (1), 60–65.

Cabaniss, S. E. 1991 Theory of variable-angle synchronous fluorescence-spectra. *Analytical Chemistry* **63** (13), 1323–1327.

Carstea, E. M., Baker, A., Bieroza, M., Reynolds, D. M. & Bridgeman, J. 2014 Characterisation of dissolved organic matter fluorescence properties by PARAFAC analysis and thermal quenching. *Water Research* **61**, 152–161.

Chen, J., LeBoef, E. J., Dai, S. & Gu, B. H. 2003a Fluorescence spectroscopic studies of natural organic matter fractions. *Chemosphere* **50** (5), 639–647.

Chen, W., Westerhoff, P., Leenheer, J. A. & Booksh, K. 2003b Fluorescence excitation – emission matrix regional integration to quantify spectra for dissolved organic matter. *Environmental Science & Technology* **37** (24), 5701–5710.

Chen, S., Yu, Y.-L. & Wang, J.-H. 2018 Inner filter effect-based fluorescent sensing systems: a review. *Analytica Chimica Acta* **999**, 13–26.

Chen, Z. Q., Li, D. & Wen, Q. X. 2020 Investigation of hydrolysis acidification process during anaerobic treatment of coal gasification wastewater (CGW): evolution of dissolved organic matter and biotoxicity. *Science of the Total Environment* **723**, 137995.

Chen, S., Liu, F., Cui, R., Zhu, B. & You, X. 2022 Removal of tetracycline hydrochloride using S–g-C3N4/PTFE membrane under visible light irradiation. *Water Cycle* **3**, 8–17.

Christensen, J., Povlsen, V. T. & Sorensen, J. 2003 Application of fluorescence spectroscopy and chemometrics in the evaluation of processed cheese during storage. *Journal of Dairy Science* **86** (4), 1101–1107.

Clark, C. D., De Bruyn, W. J., Brahm, B. & Aiona, P. 2020 Optical properties of chromophoric dissolved organic matter (CDOM) and dissolved organic carbon (DOC) levels in constructed water treatment wetland systems in southern California, USA. *Chemosphere* **247**, 125906.

Coble, P. G., Green, S. A., Blough, N. V. & Gagosian, R. B. 1990 Characterization of dissolved organic-matter in the black-sea by fluorescence spectroscopy. *Nature* **348** (6300), 432–435.

Coble, P. G., Schultz, C. A. & Mopper, K. 1993 Fluorescence contouring analysis of doc intercalibration experiment samples – a comparison of techniques. *Marine Chemistry* **41** (1–3), 173–178.

da Silva, J., Machado, A., Oliveira, C. J. S. & Pinto, M. 1998 Fluorescence quenching of anthropogenic fulvic acids by Cu(II), Fe(III) and UO22 +. *Talanta* **45** (6), 1155–1165.

Determann, S., Reuter, R., Wagner, P. & Willkomm, R. 1994 Fluorescent matter in the early atlantic-ocean .1. Method of measurement and near-surface distribution. *Deep-Sea Research Part I-Oceanographic Research Papers* **41** (4), 659–675.

Elkins, K. M. & Nelson, D. J. 2002 Spectroscopic approaches to the study of the interaction of aluminum with humic substances. *Coordination Chemistry Reviews* **228** (2), 205–225.

Elliott, S., Lead, J. R. & Baker, A. 2006 Thermal quenching of fluorescence of freshwater, planktonic bacteria. *Analytica Chimica Acta* **564** (2), 219–225.

Esteves, V. I., Santos, E. B. H. & Duarte, A. C. 1999 Study of the effect of pH, salinity and DOC on fluorescence of synthetic mixtures of freshwater and marine salts. *Journal of Environmental Monitoring* **1** (3), 251–254.

Fan, Z. H., Yang, H. L., Li, S. F. & Yu, X. 2020 Tracking and analysis of DBP precursors' properties by fluorescence spectrometry of dissolved organic matter. *Chemosphere* **239**, 124790.

Foudeil, S., Hassoun, H., Lamhasni, T., Ait Lyazidi, S., Benyaich, F., Haddad, M., Choukrad, M., Boughdad, A., Bounakhla, M., Bounouira, H., Duarte, R. M. B. O., Cachada, A. & Duarte, A. C. 2015 Catalog of total excitation-emission and total synchronous fluorescence maps with synchronous fluorescence spectra of homologated fluorescent pesticides in large use in Morocco: development of a spectrometric low cost and direct analysis as an alert method in case of massive contamination of soils and waters by fluorescent pesticides. *Environmental Science and Pollution Research* **22** (9), 6766–6777.

Gadad, P. & Nanny, M. A. 2008 Influence of cations on noncovalent interactions between 6-propionyl-2-dimethylaminonaphthalene (PRODAN) and dissolved fulvic and humic acids. *Water Research* **42** (19), 4818–4826.

Goffin, A., Vasquez-Vergara, L. A., Guerin-Rechdaoui, S., Rocher, V. & Varrault, G. 2020 Temperature, turbidity, and the inner filter effect correction methodology for analyzing fluorescent dissolved organic matter in urban sewage. *Environmental Science and Pollution Research* **27** (28), 35712–35723.

Hambly, A. C., Arvin, E., Pedersen, L. F., Pedersen, P. B., Seredynska-Sobecka, B. & Stedmon, C. A. 2015 Characterising organic matter in recirculating aquaculture systems with fluorescence EEM spectroscopy. *Water Research* **83**, 112–120.

Hanh, N. N., Durrieu, C. & Canh, T. M. 2009 Synchronous-scan fluorescence of algal cells for toxicity assessment of heavy metals and herbicides. *Ecotoxicology and Environmental Safety* **72** (2), 316–320.

He, X. S. & Fan, Q. D. 2016 Investigating the effect of landfill leachates on the characteristics of dissolved organic matter in groundwater using excitation-emission matrix fluorescence spectra coupled with fluorescence regional integration and self-organizing map. *Environmental Science and Pollution Research* **23** (21), 21229–21237.

Henderson, R. K., Baker, A., Murphy, K. R., Hamblya, A., Stuetz, R. M. & Khan, S. J. 2009 Fluorescence as a potential monitoring tool for recycled water systems: a review. *Water Research* **43** (4), 863–881.

Huang, N., Shao, W. T., Wang, W. L., Wang, Q., Chen, Z. Q., Wu, Q. Y. & Hu, H. Y. 2022 Removal of methylisothiazolinone biocide from wastewater by VUV/UV advanced oxidation process: kinetics, mechanisms and toxicity. *Journal of Environmental Management* **315**, 115107.

Huguet, A., Vacher, L., Relexans, S., Saubusse, S., Froidefond, J. M. & Parlanti, E. 2009 Properties of fluorescent dissolved organic matter in the gironde estuary. *Organic Geochemistry* **40** (6), 706–719.

Inman, E. L. & Winefordner, J. D. 1982 Constant-energy synchronous fluorescence for reduction of Raman scatter interference. *Analytica Chimica Acta* **138**, 245–252.

Islam, A., Sun, G. X., Shang, W., Zheng, X. C., Li, P. F., Yang, M. & Zhang, Y. 2021 Separation and characterization of refractory colored dissolved effluent organic matter in a full-scale industrial park wastewater treatment plant. *Environmental Science and Pollution Research* **28** (31), 42387–42400.

Ji, M. C., Li, S. J., Zhang, J. Q., Di, H., Li, F. X. & Feng, T. J. 2018 The human health assessment to phthalate acid esters (PAEs) and potential probability prediction by chromophoric dissolved organic matter EEM-FRI fluorescence in Erlong lake. *International Journal of Environmental Research and Public Health* **15** (6), 1109.

Karstens, T. & Kobs, K. 1980 Rhodamine-B and rhodamine-101 as reference substances for fluorescence quantum yield measurements. *Journal of Physical Chemistry* **84** (14), 1871–1872.

Khan, M. F. S., Akbar, M., Wu, J. & Xu, Z. 2022 A review on fluorescence spectroscopic analysis of water and wastewater. *Methods and Applications in Fluorescence* **10** (1), 012001.

Kumke, M. U., Zwiener, C., Abbt-Braun, G. & Frimmel, F. H. 1999 Spectroscopic characterization of fulvic acid fractions of a contaminated groundwater. *Acta Hydrochimica Et Hydrobiologica* **27** (6), 409–415.

Lakowicz, J. R. 2006 Plasmonics in biology and plasmon-controlled fluorescence. *Plasmonics* **1** (1), 5–33.

Lee, E. J., Yoo, G. Y., Jeong, Y., Kim, K. U., Park, J. H. & Oh, N. H. 2015 Comparison of UV-VIS and FDOM sensors for in situ monitoring of stream DOC concentrations. *Biogeosciences* **12** (10), 3109–3118.

Li, X. F., Cullen, W. R., Reimer, K. J. & Le, X. C. 1996 Microbial degradation of pyrene and characterization of a metabolite. *Science of the Total Environment* **177**, 17–29.

Li, W. T., Chen, S. Y., Xu, Z. X., Li, Y., Shuang, C. D. & Li, A. M. 2014 Characterization of dissolved organic matter in municipal wastewater using fluorescence PARAFAC analysis and chromatography multi-excitation/emission scan: a comparative study. *Environmental Science & Technology* **48** (5), 2603–2609.

Li, L., Wang, Y., Zhang, W., Yu, S., Wang, X. & Gao, N. 2020 New advances in fluorescence excitation-emission matrix spectroscopy for the characterization of dissolved organic matter in drinking water treatment: a review. *Chemical Engineering Journal* **381**, 122676.

Lin, T., Chen, H., Ding, S. X., Chen, W. & Xu, H. 2019 The fates of aromatic protein and soluble microbial product-like organics, as the precursors of dichloroacetonitrile and dichloroacetamide, in drinking water advanced treatment processes. *Environmental Science-Water Research & Technology* **5** (8), 1478–1488.

Liu, Y. C., Duan, J. M., Li, W., Beecham, S. & Mulcahy, D. 2016 Effects of organic matter removal from a wastewater secondary effluent by aluminum sulfate coagulation on haloacetic acids formation. *Environmental Engineering Science* **33** (7), 484–493.

Liu, X., Ren, Z., Ngo, H. H., He, X., Desmond, P. & Ding, A. 2021 Membrane technology for rainwater treatment and reuse: a mini review. *Water Cycle* **2**, 51–63.

Lloyd, J. B. F. 1971 Synchronized excitation of fluorescence emission spectra. *Nature-Physical Science* **231** (20), 64–65.

Lombardi, A. T. & Jardim, W. F. 1999 Fluorescence spectroscopy of high performance liquid chromatography fractionated marine and terrestrial organic materials. *Water Research* **33** (2), 512–520.

Lu, X. Q. & Jaffe, R. 2001 Interaction between Hg(II) and natural dissolved organic matter: a fluorescence spectroscopy based study. *Water Research* **35** (7), 1793–1803.

Ma, C. Q., Chen, G. Q., Wei, B. L., Shi, Y. P., Gu, L., Gao, S. M. & Zhu, T. 2011 Study on saltation of fluorescence excitation spectra of erythrosine. *Spectroscopy and Spectral Analysis* **31** (4), 1065–1068.

Mao, Y., Chen, X.-W., Chen, Z., Chen, G.-Q., Lu, Y., Wu, Y.-H. & Hu, H.-Y. 2021 Characterization of bacterial fluorescence: insight into rapid detection of bacteria in water. *Water Reuse* **11** (4), 621–631.

Maqbool, T., Zhang, J. X., Qin, Y. L., Ly, Q. V., Asif, M. B., Zhang, X. H. & Zhang, Z. H. 2020 Seasonal occurrence of N-nitrosamines and their association with dissolved organic matter in full-scale drinking water systems: determination by LC-MS and EEM-PARAFAC. *Water Research* **183**, 116096.

Mataga, N., Kaifu, Y. & Koizumi, M. 1955 The solvent effect on fluorescence spectrum – change of solute-solvent interaction during the lifetime of excited solute molecule. *Bulletin of the Chemical Society of Japan* **28** (9), 690–691.

McKnight, D. M., Boyer, E. W., Westerhoff, P. K., Doran, P. T., Kulbe, T. & Andersen, D. T. 2001 Spectrofluorometric characterization of dissolved organic matter for indication of precursor organic material and aromaticity. *Limnology and Oceanography* **46** (1), 38–48.

Mobed, J. J., Hemmingsen, S. L., Autry, J. L. & McGown, L. B. 1996 Fluorescence characterization of IHSS humic substances: total luminescence spectra with absorbance correction. *Environmental Science & Technology* **30** (10), 3061–3065.

Morling, K., Herzsprung, P. & Kamjunke, N. 2017 Discharge determines production of, decomposition of and quality changes in dissolved organic carbon in pre-dams of drinking water reservoirs. *Science of the Total Environment* **577**, 329–339.

Murphy, K. R., Stedmon, C. A., Graeber, D. & Bro, R. 2013 Fluorescence spectroscopy and multi-way techniques. PARAFAC. *Analytical Methods* **5** (23), 6557–6566.

Noomnarm, U. & Clegg, R. M. 2009 Fluorescence lifetimes: fundamentals and interpretations. *Photosynthesis Research* **101** (2–3), 181–194.

Ohno, T. 2002 Fluorescence inner-filtering correction for determining the humification index of dissolved organic matter. *Environmental Science & Technology* **36** (4), 742–746.

Pan, B., Han, X., Wu, M., Peng, H., Zhang, D., Li, H. & Xing, B. 2012 Temperature dependence of ofloxacin fluorescence quenching and complexation by Cu(II). *Environmental Pollution* **171**, 168–173.

Panigrahi, S. K. & Mishra, A. K. 2019 Inner filter effect in fluorescence spectroscopy: as a problem and as a solution. *Journal of Photochemistry and Photobiology C-Photochemistry Reviews* **41**, 100318.

Park, H. R., Kim, T. H. & Bark, K. M. 2002 Physicochemical properties of quinolone antibiotics in various environments. *European Journal of Medicinal Chemistry* **37** (6), 443–460.

Patel-Sorrentino, N., Mounier, S. & Benaim, J. Y. 2002 Excitation-emission fluorescence matrix to study pH influence on organic matter fluorescence in the Amazon basin rivers. *Water Research* **36** (10), 2571–2581.

Peuravuori, J., Koivikko, R. & Pihlaja, K. 2002 Characterization. differentiation and classification of aquatic humic matter separated with different sorbents: synchronous scanning fluorescence spectroscopy. *Water Research* **36** (18), 4552–4562.

Provenzano, M. R., D'Orazio, V., Jerzykiewicz, M. & Senesi, N. 2004 Fluorescence behaviour of Zn and Ni complexes of humic acids from different sources. *Chemosphere* **55** (6), 885–892.

Pulgarin, J. A. M. & Molina, A. A. 1994 Matrix isopotential synchronous fluorescence direct determination of gentisic acid in urine. *Analytica Chimica Acta* **296** (1), 87–97.

Pullin, M. J. & Cabaniss, S. E. 1995 Rank analysis of the ph-dependent synchronous fluorescence-spectra of six standard humic substances. *Environmental Science & Technology* **29** (6), 1460–1467.

Qiu, T., Xu, X. Y. & Qian, X. H. 2010 Solvent effects for fluorescence and absorption of tetra (fluoroalkyl) metallophthalocyanines: fluorocarbon solvent cage. *Journal of Photochemistry and Photobiology A-Chemistry* **214** (1), 86–91.

Rechthaler, K. & Kohler, G. 1994 Excited-state properties and deactivation pathways of 7-aminocoumarins. *Chemical Physics* **189** (1), 99–116.

Ren, L., Liu, C., Meng, T. & Sun, Y. 2021 Effects of micro-flocculation pretreatment on the ultrafiltration membrane fouling caused by different dissolved organic matters in treated wastewater. *Water Reuse* **11** (4), 597–609.

Reynolds, D. M. 2003 Rapid and direct determination of tryptophan in water using synchronous fluorescence spectroscopy. *Water Research* **37** (13), 3055–3060.

Rinnan, A., Booksh, K. S. & Bro, R. 2005 First order Rayleigh scatter as a separate component in the decomposition of fluorescence landscapes. *Analytica Chimica Acta* **537** (1–2), 349–358.

Rodriguez, J. J. S. & Sanz, C. P. 2000 Fluorescence techniques for the determination of polycyclic aromatic hydrocarbons in marine environment: an overview. *Analusis* **28** (8), 710–717.

Rodriguez-Vidal, F. J., Garcia-Valverde, M., Ortega-Azabache, B., Gonzalez-Martinez, A. & Bellido-Fernandez, A. 2020 Characterization of urban and industrial wastewaters using excitation-emission matrix (EEM) fluorescence: searching for specific fingerprints. *Journal of Environmental Management* **263**, 110396.

Samokhvalov, A. 2020 Analysis of various solid samples by synchronous fluorescence spectroscopy and related methods: a review. *Talanta* **216**, 120944.

Sanchez, N. P., Skeriotis, A. T. & Miller, C. M. 2013 Assessment of dissolved organic matter fluorescence PARAFAC components before and after coagulation-filtration in a full scale water treatment plant. *Water Research* **47** (4), 1679–1690.

Sgroi, M., Roccaro, P., Korshin, G. V., Greco, V., Sciuto, S., Anumol, T., Snyder, S. A. & Vagliasindi, F. G. A. 2017 Use of fluorescence EEM to monitor the removal of emerging contaminants in full scale wastewater treatment plants. *Journal of Hazardous Materials* **323**, 367–376.

Shen, H., Tang, X. C., Wu, N. X. & Chen, H. B. 2018 Leakage of soluble microbial products from biological activated carbon filtration in drinking water treatment plants and its influence on health risks. *Chemosphere* **202**, 626–636.

Sheng, G. P. & Yu, H. Q. 2006 Characterization of extracellular polymeric substances of aerobic and anaerobic sludge using three-dimensional excitation and emission matrix fluorescence spectroscopy. *Water Research* **40** (6), 1233–1239.

Shi, W., Zhuang, W.-E., Hur, J. & Yang, L. 2021 Monitoring dissolved organic matter in wastewater and drinking water treatments using spectroscopic analysis and ultra-high resolution mass spectrometry. *Water Research* **188**, 116406.

Siqintuya, Sueishi, Y. & Yamamoto, S. 2005 Polar and hydrogen-bonding effects of alcohols on the emission spectrum of styrene-triethylamine system. *Journal of Solution Chemistry* **34** (10), 1109–1119.

Stedmon, C. A. & Bro, R. 2008 Characterizing dissolved organic matter fluorescence with parallel factor analysis: a tutorial. *Limnology and Oceanography-Methods* **6**, 572–579.

Stedmon, C. A., Markager, S. & Bro, R. 2003 Tracing dissolved organic matter in aquatic environments using a new approach to fluorescence spectroscopy. *Marine Chemistry* **82** (3–4), 239–254.

Stuhlmann, B., Grassle, A. & Schmitt, M. 2014 Determination of the geometry change of 5-cyanoindole upon electronic excitation from a combined franck-Condon/rotational constants fit. *Physical Chemistry Chemical Physics* **16** (3), 899–905.

Tan, Y. W., Lin, T., Jiang, F. C., Dong, J., Chen, W. & Zhou, D. J. 2017 The shadow of dichloroacetonitrile (DCAN), a typical nitrogenous disinfection by-product (N-DBP), in the waterworks and its backwash water reuse. *Chemosphere* **181**, 569–578.

Trueman, B. F., MacIsaac, S. A., Stoddart, A. K. & Gagnon, G. A. 2016 Prediction of disinfection by-product formation in drinking water via fluorescence spectroscopy. *Environmental Science-Water Research & Technology* **2** (2), 383–389.

van Eck, N. J. & Waltman, L. 2010 Software survey: VOSviewer, a computer program for bibliometric mapping. *Scientometrics* **84** (2), 523–538.

Vera, M., Martin-Alonso, J., Mesa, J., Granados, M., Beltran, J. L., Casas, S., Gibert, O. & Cortina, J. L. 2017 Monitoring UF membrane performance treating surface-groundwater blends: limitations of FEEM-PARAFAC on the assessment of the organic matter role. *Chemical Engineering Journal* **317**, 961–971.

Vo-Dinh, T. 1978 Multicomponent analysis by synchronous luminescence spectrometry. *Analytical Chemistry* **50** (3), 396–401.

Wakebe, T. & Van Keuren, E. 1999 The excitation spectra of two-photon induced fluorescence in xanthene dyes. *Japanese Journal of Applied Physics Part 1-Regular Papers Brief Communications & Review Papers* **38** (6A), 3556–3561.

Wang, C., Yuan, Z.-X., Liu, Y.-Y., Wu, Q.-Y. & Sun, Y.-X. 2021 Relative developmental toxicities of reclaimed water to zebrafish embryos and the relationship with relevant water quality parameters. *Water Cycle* **2**, 85–90.

Wang, S., Li, Y., Xiao, K. & Huang, X. 2022 Fluorescence excitation-emission matrix as a novel indicator of assimilable organic carbon in wastewater: implication from a coal chemical wastewater study. *Science of the Total Environment* **804**, 150144.

Watras, C. J., Hanson, P. C., Stacy, T. L., Morrison, M., Mather, J., Hu, Y. H. & Milewski, P. 2011 A temperature compensation method for CDOM fluorescence sensors in freshwater. *Limnology and Oceanography-Methods* **9**, 296–301.

Wei, Z. M., Xi, B. D., Wang, S. P., Xu, J. G., Zhou, Y. Y. & Liu, H. H. 2005 Fluorescence characteristic changes of dissolved organic matter during municipal solid waste composting. *Journal of Environmental Sciences* **17** (6), 953–956.

Westerhoff, P., Chen, W. & Esparza, M. 2001 Fluorescence analysis of a standard fulvic acid and tertiary treated wastewater. *Journal of Environmental Quality* **30** (6), 2037–2046.

Wu, J., Pons, M. N. & Potier, O. 2006 Wastewater fingerprinting by UV-visible and synchronous fluorescence spectroscopy. *Water Science and Technology* **53** (4–5), 449–456.

Xie, G. B., Sueishi, Y. & Yamamoto, S. 2004 Effects of alcohols on emission spectra of toluene-triethylamine mixtures in THF: separation into polar and hydrogen-bonding interactions. *Journal of Photochemistry and Photobiology A-Chemistry* **162** (2–3), 449–456.

Xu, A., Wu, Y.-H., Chen, Z., Wu, G., Wu, Q., Ling, F. Huang, W. & Hu, H.-Y. 2020 Towards the new era of wastewater treatment of China: development history, current status, and future directions. *Water Cycle* **1**, 80–87.

Xu, X. T., Kang, J., Shen, J. M., Zhao, S. X., Wang, B. Y., Zhang, X. X. & Chen, Z. L. 2021 EEM-PARAFAC characterization of dissolved organic matter and its relationship with disinfection by-products formation potential in drinking water sources of northeastern China. *Science of the Total Environment* **774**, 145297.

Yadav, M. K., Aryal, R., Short, M. D. & Saint, C. P. 2019 Fluorescence excitation-emission spectroscopy: an analytical technique to monitor drugs of addiction in wastewater. *Water* **11** (2), 377.

Yamashita, Y. & Jaffe, R. 2008 Characterizing the interactions between trace metals and dissolved organic matter using excitation-emission matrix and parallel factor analysis. *Environmental Science & Technology* **42** (19), 7374–7379.

Yang, X.-P., Shi, B.-F., Zhang, Y.-H., Tang, J. & Cai, D.-C. 2008 Identification of polycyclic aromatic hydrocarbons (PAHs) in soil by constant energy synchronous fluorescence detection. *Spectrochimica Acta Part A-Molecular and Biomolecular Spectroscopy* **69** (2), 400–406.

Yang, L. Y., Hur, J. & Zhuang, W. N. 2015a Occurrence and behaviors of fluorescence EEM-PARAFAC components in drinking water and wastewater treatment systems and their applications: a review. *Environmental Science and Pollution Research* **22** (9), 6500–6510.

Yang, L. Y., Kim, D., Uzun, H., Karanfil, T. & Hur, J. 2015b Assessing trihalomethanes (THMs) and N-nitrosodimethylamine (NDMA) formation potentials in drinking water treatment plants using fluorescence spectroscopy and parallel factor analysis. *Chemosphere* **121**, 84–91.

Yang, L. Y., Chen, W., Zhuang,W. E., Cheng, Q., Li,W. X., Wang, H., Guo, W. D., Chen, C. T. A. & Liu, M. H. 2019 Characterization and bioavailability of rainwater dissolved organic matter at the southeast coast of China using absorption spectroscopy and fluorescence EEM-PARAFAC. *Estuarine Coastal and Shelf Science* **217**, 45–55.

Zhang, F., Wang, X. P., Chen, Y. & Airiken, M. 2020a Estimation of surface water quality parameters based on hyper-spectral and 3D-EEM fluorescence technologies in the Ebinur Lake Watershed, China. *Physics and Chemistry of the Earth* **118**, 102895.

Zhang, H., Cui, K. P., Guo, Z., Li, X. Y., Chen, J., Qi, Z. G. & Xu, S. Y. 2020b Spatiotemporal variations of spectral characteristics of dissolved organic matter in river flowing into a key drinking water source in China. *Science of the Total Environment* **700**, 134360.

Zsolnay, A., Baigar, E., Jimenez, M., Steinweg, B. & Saccomandi, F. 1999 Differentiating with fluorescence spectroscopy the sources of dissolved organic matter in soils subjected to drying. *Chemosphere* **38** (1), 45–50.

First received 8 October 2022; accepted in revised form 27 November 2022. Available online 13 December 2022

doi: 10.2166/wrd.2023.003

Bioremoval efficiency and metabolomic profiles of cellular responses of *Chlorella pyrenoidosa* to phenol and 4-fluorophenol

Min Li [a,b],*, Lu Ma[a], Yajing An[a], Dongbin Wei **IWA**[c], Haijun Ma[a,b], Juanjuan Zhu[a,b] and Cuiping Wang[a,b]

[a] College of Biological Science and Engineering, North Minzu University, Yinchuan 750021, Ningxia Province, People's Republic of China
[b] Ningxia Grape & Wine Innovation Center, Yinchuan, Ningxia Province 750021, People's Republic of China
[c] State Key Laboratory of Environmental Chemistry and Ecotoxicology, Research Center for Eco-Environmental Sciences, Chinese Academy of Sciences, Beijing 100085, People's Republic of China
*Corresponding author. E-mail: bkdlimin@126.com

ML, 0000-0003-4706-5672

ABSTRACT

We examined the growth tolerance, bioremoval efficacy, and metabolomic profiles of the cellular responses of *Chlorella pyrenoidosa* to phenol and 4-fluorophenol. We found that *C. pyrenoidosa* can tolerate exposure to 100 mg/L of phenol and 4-fluorophenol, and the growth of algal cells had a significant hormesis of inhibition first and then promotion. Up to 70% bioremoval of phenol and 4-fluorophenol may occur after 240 h of treatment. Superoxide dismutase (SOD) and catalase (CAT) activities, malondialdehyde (MDA) content, and reactive oxygen species (ROS) in algal cells in the phenol- and 4-fluorophenol-treated groups were similar or lower than in the control group. Furthermore, photosynthetic pigment and glycerophospholipid contents were significantly upregulated in both phenol- and 4-fluorophenol-treated groups, as indicated by the metabolomic analysis of *C. pyrenoidosa*, resulting in the vigorous growth of algal cells compared to the control group. Therefore, *C. pyrenoidosa* can be an excellent biosorbent for phenol and 4-fluorophenol.

Key words: *Chlorella pyrenoidosa*, 4-fluorophenol, metabolomics, phenol, removal

HIGHLIGHTS

- *Chlorella pyrenoidosa* can well tolerate 100 mg/L of phenol or 4-fluorophenol exposure.
- 100 mg/L phenol or 4-fluorophenol can be bioremoved by >70% after 240 h of treatment.
- Significant hormesis in algal growth was observed during the treatment.
- Growth promotion is attributed to chlorophyll and glycerophospholipid accumulation.

GRAPHICAL ABSTRACT

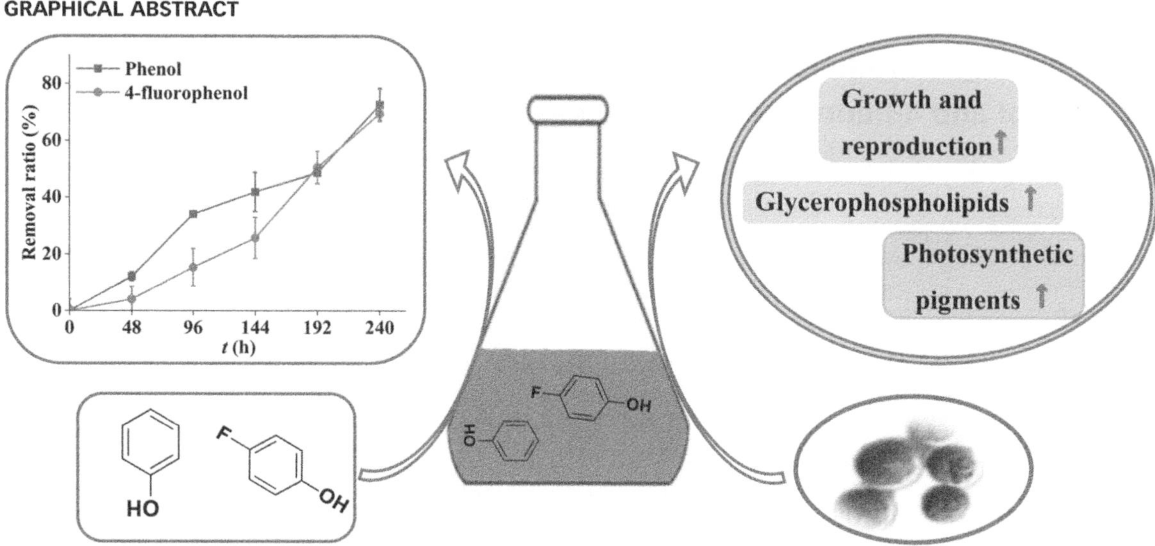

1. INTRODUCTION

Phenol and halophenols are widely used to produce pharmaceuticals, herbicides, dyes, disinfectants, and preservatives (Brycht *et al.* 2016; Fawzy & Alharthic 2021). Environmental pollution involving these compounds is directly proportional to extensive industrialization. Such substances have been detected in various water bodies with a broad range of concentrations (Wang 2022). For instance, the concentration of 4-chlorophenol reached 37.60–47.81 ng/L in the Yangtze River of China (Zhu *et al.* 2022). In contrast, it varies between 100 and 1,000 mg/L in effluents from relevant industries (Patel *et al.* 2022). Phenol concentration in wastewater varies between 10 and 300 mg/L but reaches 4.5 g/L in highly polluted ones (Al-Khalid & El-Naas 2012). Phenol and halophenols, especially halophenols with one or more halogen atoms, are resistant to degradation, persisting in the environment, and resulting in a range of environmental pollution and human health risks (Ge *et al.* 2017; Juksu *et al.* 2019). Halophenols have been categorized as priority pollutants in Europe, the USA, and China, given their hazardous nature (Duan *et al.* 2019). Consequently, the elimination of phenol and halophenol is an important requirement for eco-environmental safety.

Various physicochemical and biological processes have been successfully explored to remove phenol and halophenols. Biological treatment is eco-friendly due to its minimal risk of toxic secondary products compared with physicochemical techniques. Microalgae-mediated bioremediation is highly preferred as a candidate for pollutant removal (Ansar *et al.* 2022; Couto *et al.* 2022). Microalgae are nontargets for antibiotics (Escudero *et al.* 2020; Sawaya *et al.* 2022). Microalgae can be reused as fuel, pigments, fertilizers, and medicine according to the pollutant removal mechanism after proper treatments (Andreotti *et al.* 2020; Jalilian *et al.* 2020; Ong *et al.* 2021; Reddy *et al.* 2021). However, phenol and halophenol treatment by algae has been studied far less than that of fungi and bacteria (Fawzy & Alharthic 2021).

Biosorption, bioaccumulation, and biodegradation are essential functions played by microalgae while removing organic pollutants (Xiong *et al.* 2016; Chen *et al.* 2021a). Biosorption is the first step in the bioaccumulation process, where pollutants are absorbed into cells by metabolism-dependent active transport systems (Novák *et al.* 2020). The interaction between algal cells and pollutants is complex, as efficient pollutant removal is affected by potential adverse effects on algal cells. Biosorption is challenging when pollutant concentrations are high. In such cases, photosynthesis decreases and lipid peroxidation is produced as a result of pollution stress caused by pollutants, leading algae cells to irreversible injuries or death (Chin *et al.* 2019; Chen *et al.* 2021b; Fawzy & Alharthic 2021; Zhai *et al.* 2022). Therefore, it is necessary to analyse the growth and physiological properties of algal cells when exposed to pollutants and their removal efficacy.

Algal-based pollutant removal usually depends on the substance structure and the algal species (Xiao *et al.* 2021). *Chlorella pyrenoidosa* is one of the most representative green algae and is associated with strong adaptability, rapid reproduction, and sensitivity to pollutants. Phenol and 4-fluororphenol are widely used. Consequently, the availability of these substances increases in the environment. Phenol and 4-fluororphenol differ in structure, toxicity, and biodegradability. In this study, the removal efficacy of phenol and 4-fluororphenol by *C. pyrenoidosa* was studied. The effect of phenol and 4-fluorophenol on the growth rate, antioxidant enzyme activity, and metabonomics of *C. pyrenoidosa* was explored. The results are beneficial for examining the remediation potential and mechanism of phenol/4-fluorophenol wastewater by microalgae.

2. MATERIALS AND METHODS

2.1. *C. pyrenoidosa* growth inhibition test

C. pyrenoidosa (collection no. FACHB-5) was purchased from the Freshwater Algae Culture Collection at the Institute of Hydrobiology, Chinese Academy of Sciences, Wuhan, China. Algal cells were grown in a 250-mL flask containing 100 mL of BG11 medium. All flasks were kept in an artificial climate box with a $25 \pm 0.2\,°C$, 12:12 h light (4,000 lux)-dark cycle. All flasks were shaken manually three times a day.

Phenol and 4-fluorophenol were dissolved in methanol and the volume of methanol in the exposure solution was always $\leq 0.5\%$. Solutions of 0.5% and non-methanol were set as the solvents and blank controls. A t-test was conducted between those two controls and $p > 0.05$ was obtained, indicating that methanol toxicity could be negligible under this threshold. The initial algal density was adjusted to $(1–2) \times 10^5$ cells/L in the exposure solutions. The initial concentrations of phenol and 4-fluorophenol were 0, 10, 50, and 100 mg/L. An algal density (ρ, cells/mL) measurement was conducted by haemocytometer every 24 h until 240 h and the inhibition ratio $(\%) = (\rho_{control}–\rho_{treatment})/\rho_{control}$ was calculated (OECD 2006).

2.2. Determination of phenol and 4-fluorophenol

The phenol and 4-fluorophenol concentrations in the exposure solution were determined using high-performance liquid chromatography (HPLC, Agilent 1200, ZORBAX RX-C18) (4.6 mm × 150 mm, 5 μm) with a diode array detector (testing λ was 270 nm). The removal ratio (%) of phenol/4-fluorophenol was calculated according to the ratio of its removal concentration to its initial concentration. Additionally, molecular spectrum scanning (λ from 250 to 400 nm) of phenol/4-fluorophenol cultured in the BG11 medium was conducted every 24 h until 240 h to examine the abiotic degradation of phenol/4-fluorophenol.

2.3. Determination of physiological and biochemical indexes

Ten mL of algal solution was centrifuged at 4,000 rpm for 10 min. Next, the algal pellet was suspended in 10 mL phosphate buffer saline (PBS, 0.01 mol/L, pH 7.8). Subsequently, the solution was ultrasonicated on ice for 15 min at 200 W, ultrasonic time: 3 s, and rest time: 7 s to break cells. Later, the crude enzyme solution was obtained by centrifugation at 4,000 rpm. The total soluble protein (TSP), superoxide dismutase (SOD), catalase (CAT), malondialdehyde (MDA), and reactive oxygen species (ROS) contents were measured using a commercial kit (Keming, Suzhou, China).

2.4. Metabolomics

Algal cells treated with 100 mg/L phenol/4-fluorophenol for 96 h were collected for metabolomic analysis. Briefly, 20 mL of samples were ultrasound-crushed, freeze-dried, and ultrasonically extracted with 10 mL of methanol. The extracted liquid was centrifuged (10,000 g for 10 min), filtered (0.2 μm filter membrane) and used for metabolite detection. An Agilent 6546 UPLC-Q/TOF-MS containing a ZORBAX Eclipse Plus C18 (2.1 × 50 mm, 1.8 μm) was used to run the samples. The mobile phases were (A) H_2O with 0.1% ammonium acetate and (B) methanol. Gradient elution was performed at a flow rate of 0.4 mL/min under the following programme: held at 10% B from 0 to 3 min; then, B changed linearly from 10 to 70% in the next 3–15 min. Over the next 0.10 min, B changed to 100% and was held for 3 min. The column temperature was kept at 30 °C. The MS injection volume for each sample was 2 μL. The operating parameters for the Q/TOF-MS in both positive and negative ion modes were: gas temperature 325 °C, drying gas flow 10 L/min, nebulizer pressure 40 psi, mass scan range from 50 to 1,500 m/z, and acquisition rate of 200 ms/spectrum.

The raw data were converted into CEF format using mass hunter profinder software (Agilent Corp., 10.1) and imported into Mass Profiler Professional software (version 15.1) for further analysis. A one-way analysis of variance (ANOVA) was used to determine the differences among the control, phenol, and 4-fluorophenol-treated groups. A t-test was used to determine the differences between the control vs. phenol-treated groups and the control vs. 4-fluorophenol-treated groups. A significant difference occurred when $p < 0.05$. Unsupervised principal component analysis (PCA) and orthogonal partial least squares-discriminant analysis (OPLS-DA) were applied to obtain an overview of the chemical variations among treatments. METLIN PCDL B.08.00 was used to identify the compounds. The differential metabolites were mapped to the Kyoto Encyclopedia of Genes and Genomes (KEGG) database and the corresponding metabolic pathways were searched.

3. RESULTS AND DISCUSSION

3.1. The tolerance and removal efficacy of *C. pyrenoidosa* to phenol and 4-fluorophenol

The tolerance of *C. pyrenoidosa* to phenol and 4-fluorophenol exposure was evaluated based on a growth inhibition test. The results are shown in Figure 1(a) and 1(b). The tolerance was significantly correlated with exposure time and concentration. Each 10 mg/L of phenol and 4-fluorophenol had a weak promoting effect on the growth of algal cells, while the first inhibited and then promoted growth trend was observed in the culture of each 100 mg/L of phenol and 4-fluorophenol. The most significant promotional effect of phenol and 4-fluorophenol on the growth of algal cells was observed at 96 h. The above results can be summarized as the U-shaped effect found in many microalgae after being stimulated by exogenous substances. Microalgae can maintain the vitality and stability of the algal system by increasing the number of algal cells after stimulation by microplastics (Li *et al.* 2022). Namely, the increase in algal density could be an apparent stress-resistance response of the microalgae.

Phenol and 4-fluorophenol had good stability in the BG11 medium in the 240 h tested (Supplementary material, Figure S1). However, within the algal cells, it was found that 10 mg/L phenol could be removed in 144 h. In contrast, the removal ratios of 50 and 100 mg/L phenol after 240 h of exposure were 85.23 and 72.64%, respectively (Figure 1(c)). The removal efficiency of 4-fluorophenol by algal cells was slightly lower than that of phenol, with removal ratios of 10, 50, and 100 mg/L 4-fluorophenol after 240 h of exposure of 91.92, 80.46, and 70.51%, respectively (Figure 1(d)). Compared with phenol, the tolerance and removal efficacy of 4-fluorophenol by *C. pyrenoidosa* decreased slightly in this study.

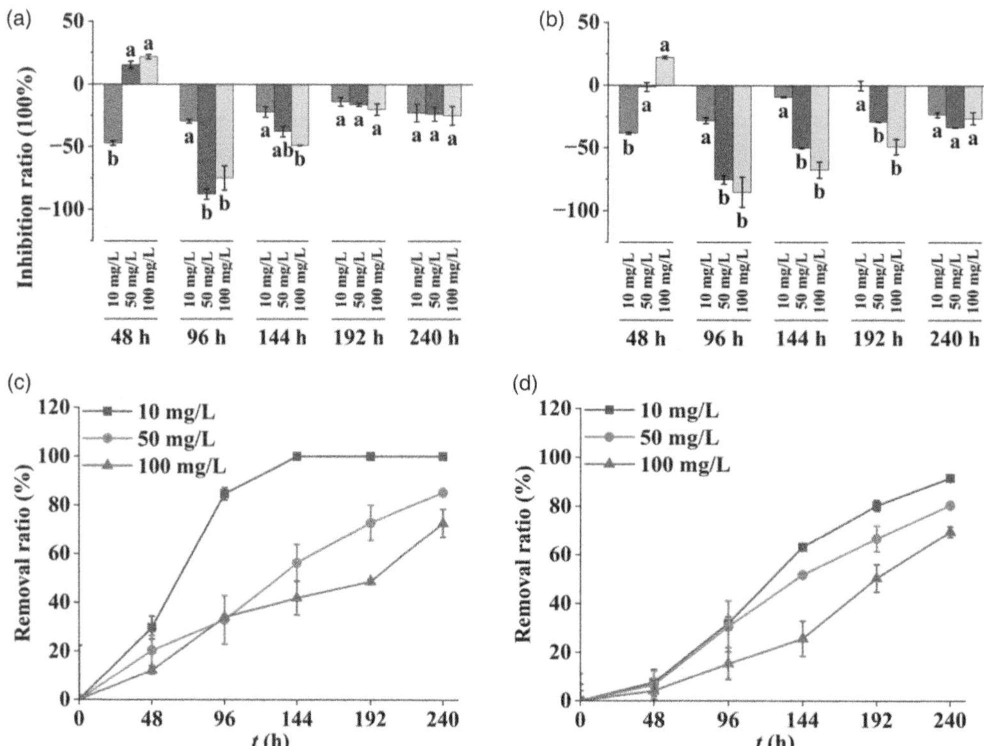

Figure 1 | The interactions of *Chlorella pyrenoidosa* with phenol and 4-fluorophenol: (a) the growth inhibition effect of phenol on *C. pyrenoidosa*; (b) the growth inhibition effect of 4-fluorophenol on *C. pyrenoidosa*; (c) the removal efficacy of *C. pyrenoidosa* on phenol; and (d) the removal efficacy of *C. pyrenoidosa* on 4-fluorophenol.

3.2. The effect of phenol and 4-fluorophenol on *C. pyrenoidosa* ROS production and oxidative stress

The effects of phenol and 4-fluorophenol on the ROS and antioxidant enzymes of algal cells are shown in Figure 2. There was a slight variation in the ROS and SOD levels observed ($Value_{sample}/Value_{control}$ is close to 1) in response to phenol and 4-fluorophenol treatments. The soluble protein (SP) contents in the phenol and 4-fluorophenol treatment groups were significantly higher than those in the control group, indicating a vigorous growth of algal cells compared to the control group. However, the CAT and MDA contents in the phenol- and 4-fluorophenol-treated groups were lower than in the control group.

ROS, which includes hydroxyl radical (OH), superoxide radical (O_2^-), hydrogen peroxide (H_2O_2), etc., is a typical product of cellular metabolism. Excess ROS can lead to oxidative damage and disturb cell metabolism. Antioxidant defence systems maintained by SOD, CAT, etc., may prevent cellular damage caused by ROS and maintain physiological homeostasis. CAT is an inducible enzyme and ROS induces its gene promoter. However, the production of inducible CAT always lags behind ROS production. Therefore, the oxidation caused by ROS cannot be eliminated by CAT, leading to the production of MDA. The low levels of ROS, CAT, and MDA in the intracellular algal cells in the phenol and 4-fluorophenol treatment groups were stable and healthy compared with the control group. This is consistent with the results of the growth inhibition test, in which it was shown that the growth of algal cells after 96 h of treatment could be significantly promoted by phenol and 4-fluorophenol.

3.3. The effects of phenol and 4-fluorophenol on the metabolomics of *C. pyrenoidosa*

The metabolites of *C. pyrenoidosa* were investigated to further explore the effects of phenol and 4-fluorophenol on the growth of algal cells. Thirty-eight upregulated and 22 downregulated metabolites were screened in the phenol-treated group using *t*-test analysis. The conditions of a fold change of <-2 or >2 and a *p*-value of <0.05 were observed. At the same time, 34 upregulated metabolites and 12 downregulated metabolites in the 4-fluorophenol-treated group were screened. The clustering analysis of these metabolites is shown in Figure 3(a) and 3(b). It was found that there were 34 common differential metabolites between the phenol- and 4-fluorophenol-treated groups. More metabolites were upregulated than downregulated in phenol or 4-fluorophenol, indicating that most physiological activities were stimulated in *C. pyrenoidosa*. Differential metabolites of the phenol- and 4-fluorophenol-treated groups were used in metabolic pathway enrichment analysis and the results of this analysis are shown in Figure 3(c) and 3(d). Algae in the phenol- and 4-fluorophenol-treated groups were significantly affected regarding porphyrin and glycerophospholipid metabolisms.

Generally, the generation of photosynthetic pigments is closely related to microalgal growth (Liu *et al.* 2021). Photosynthetic pigments are the material basis for photosynthesis in plants and microalgae, and their content is considered a sensitive parameter under environmental stress conditions (Qian *et al.* 2018). As the most abundant universal chlorophyll pigment in plants, algae, and cyanobacteria, Chl-a is a valuable parameter that reflects the efficiency of light absorption by algal cells (Sutherland *et al.* 2015), and its biosynthesis is heterogeneous (Ioannides *et al.* 1994). The biosynthesis of chlorophyll in algal cells occurs via multiple and parallel biosynthetic routes and results in the formation and accumulation of various intermediate or transform products. Porphobilinogen is a pyrrole derivative essential for magnesium protoporphyrin synthesis. Chlorophyllide-a is the biosynthetic precursor of chlorophyll-a. 7-Hydroxychlorophyllide is an intermediate

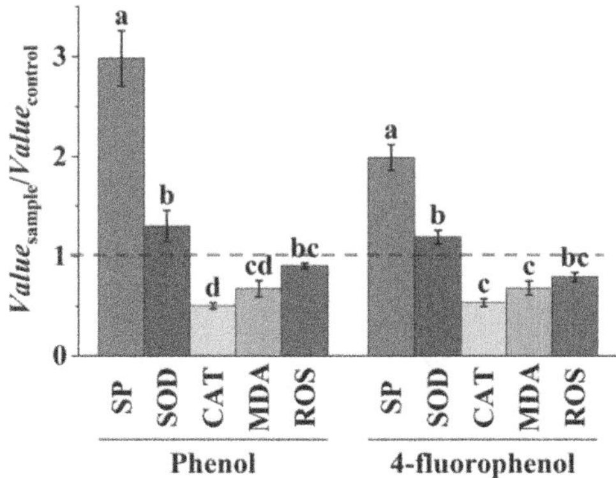

Figure 2 | Antioxidant enzyme activity and ROS levels of *Chlorella pyrenoidosa* treated for 96 h with 100 mg/L of phenol and 4-fluorophenol. SP, soluble protein.

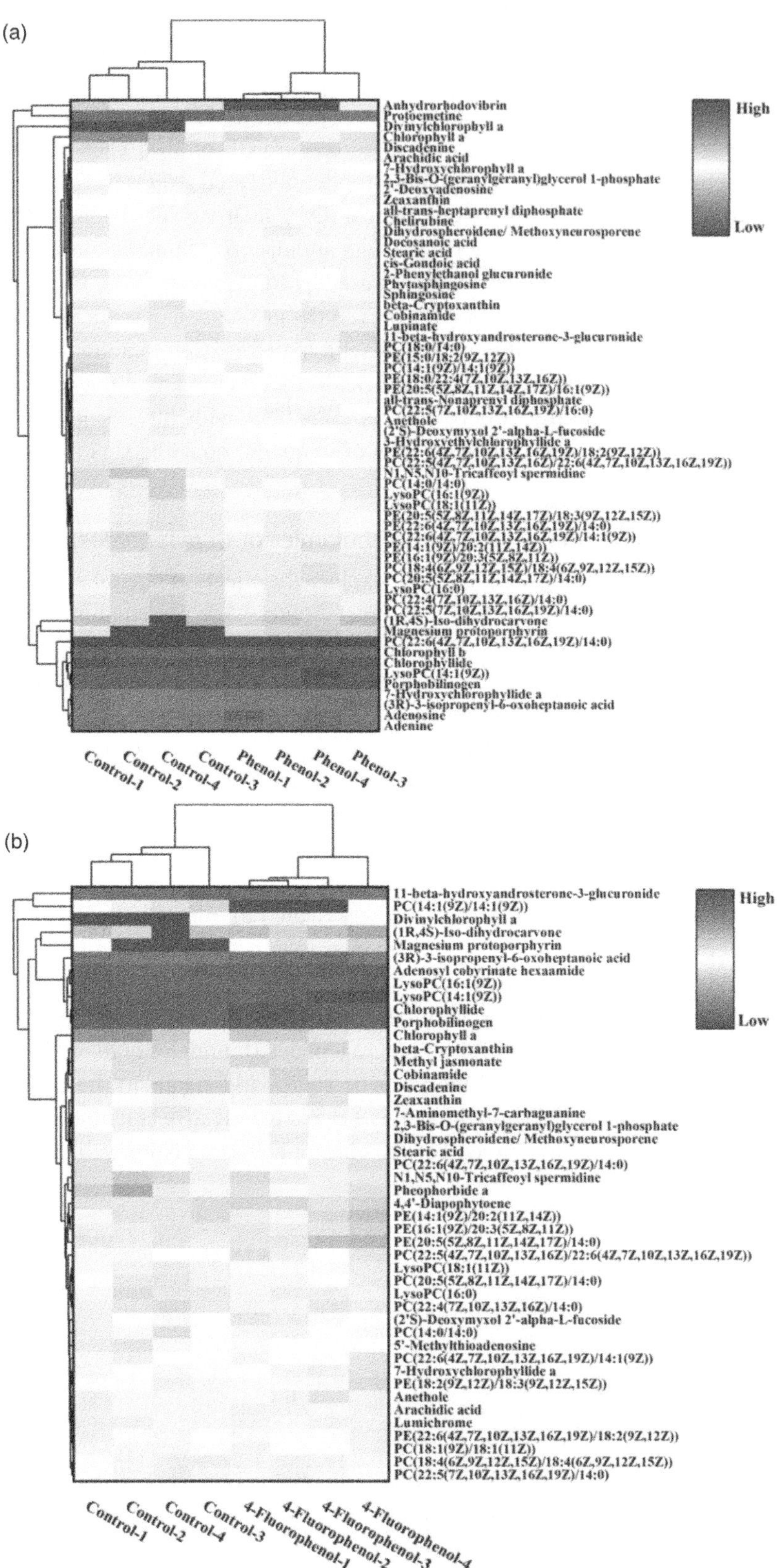

Figure 3 | Metabolomic analysis of *Chlorella pyrenoidosa* treated with each 100 mg/L of phenol and 4-fluorophenol for 96 h: (a) heat map of the differential metabolites of the phenol-treated group; (b) heat map of the differential metabolites of the 4-flououphenol-treated group; (c) bubble diagram of metabolic pathway analysis of the phenol-treated group; and (d) bubble diagram of metabolic pathway analysis of 4-fluorophenol-treated group. (*continued.*).

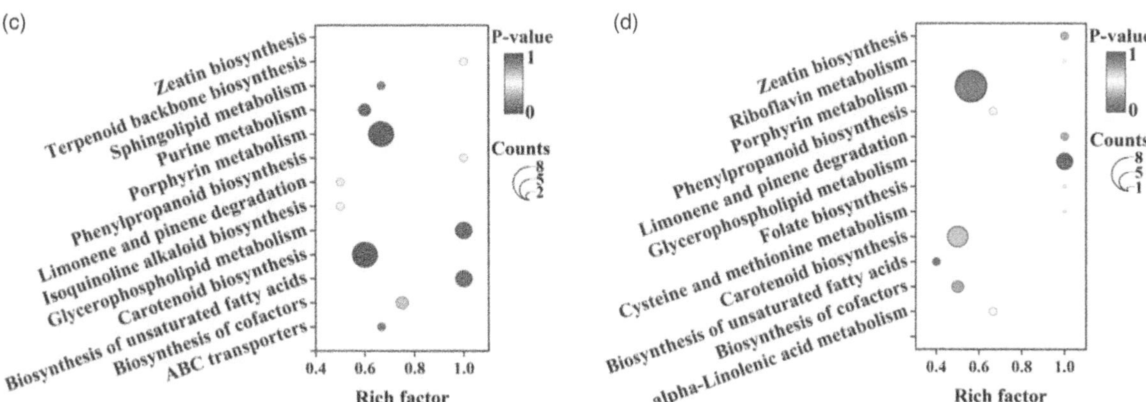

Figure 3 | Continued.

product of the transformation of chlorophylls a and b (Yang *et al.* 2016). Divinyl chlorophyll a and b are significant chlorophylls in the euphotic zone of tropical waters (Rebeiz *et al.* 1994).

Porphobilinogen, magnesium protoporphyrin, chlorophyllide, chlorophyll a, divinyl chlorophyll a, and 7-hydroxychlorophyllide a were upregulated in the phenol- and 4-fluororphenol-treated groups and chlorophyll b, 3-hydroxyethyl chlorophyllide a were upregulated in the phenol-treated group compared to the control group. In contrast, pheophorbide a was upregulated in the 4-fluorophenol-treated group. All these upregulated metabolites indicated that phenol and 4-flourophenol could significantly induce the biosynthesis of photosynthetic pigments, leading to the growth promotion of algal cells. It was reported that pharmaceuticals and personal care products (PPCPs) like ibuprofen, tylosin, erythromycin, and sulfamethoxazole could increase the content of chlorophyll a in algal cells, and the phenomenon was possibly attributed to compounds containing nitrogen atoms that can provide a nitrogen source or hormesis effect (Pinckney *et al.* 2013; Ding *et al.* 2017; Zhang *et al.* 2020). Considering that there is no nitrogen in phenol or 4-fluorophenol formulas, the hormesis effect may be the possible reason for the upregulation of chlorophyll a and its related compounds in the treated groups in this study.

Lipids are involved in regulating the response of organisms to external stress (Upchurch 2008; Chen *et al.* 2021c). For algal cells, membrane components (e.g., glycerophospholipids, betaine ether lipids, glycosyl glycerides, and phosphoglycerides) and storage lipids (mainly triacylglycerols) are the major lipid classes (Li-Beisson *et al.* 2019). There were 10 kinds of phosphatidylcholine (PC), eight kinds of phosphatidylethanolamine (PE), and four kinds of lysophosphatidylcholine (lysoPC) screened in this study as differential metabolites in the phenol-treated group, and all of these glycerophospholipids were upregulated. Similarly, there were 19 differential glycerophospholipid metabolites (10 kinds of PC, five kinds of PE, and four kinds of lysoPC), and 17 of the 19 compounds were upregulated in the 4-fluorophenol-treated group. In general, glycerophospholipids are metabolites of the average growth of algal cells. Changes in that composition can severely alter membrane fluidity, thickness, packing, and the dynamics and functions of membrane proteins (Hu *et al.* 2015; Huang *et al.* 2015). Unsaturated fatty acids are essential components of glycerophospholipids. Due to the antioxidant capacity attributed to their electron-donating ability (Huang & Wang 2004), the increase of unsaturated fatty acids benefits cell metabolism, which likely plays a role in stimulating cell growth.

Due to hydrogen and fluorine atoms being sterically quite similar, with their Vander walls radii being 1.2 and 1.35 Å, fluorine is one of the most classical bioisosteric replacements of hydrogen (Lima & Barreiro 2005). However, the electronegativity of fluorine is greater than that of hydrogen. Hydrogen replacement with fluorine could alter the physicochemical and biological activity of organic compounds. Each 100 mg/L of phenol and 4-fluorophenol can significantly stimulate the intracellular metabolism of *C. pyrenoidosa* in this study. The upregulation of photosynthetic pigments and glycerol phospholipids is the common essential mechanism of these two compounds to promote the growth of algal cells. ANOVA, combined PCA, and OPLS-DA analyses were conducted to explore the metabolite differences among all treated groups. The loading plots and cluster analysis are shown in Supplementary material, Figure S2(a)–S2(c). Most metabolites of the phenol and 4-fluorophenol-treated groups had the same change trend compared to the control group. Namely, after 100 mg/L of phenol or 4-fluorophenol treatment for 96 h, *C. pyrenoidosa* showed similar tolerance and metabolic response.

4. CONCLUSION

The tolerance of *C. pyrenoidosa* to the substrates of phenol and 4-fluorophenol had a dose-dependent and time-dependent relationship, but the difference caused by the substrates was not apparent. In general, *C. pyrenoidosa* can grow well in ≤ 100 mg/L phenol or 4-fluorophenol exposure, while the removal efficiency of phenol or 4-fluorophenol by algal cells decreases with an increase in the exposure dose. The removal efficiency of 10 or 50 mg/L phenol or 4-fluorophenol by algal cells is higher than that of 100 mg/L. Anyhow, 100 mg/L phenol or 4-fluorophenol-treated by *C. pyrenoidosa* for 240 h had more than 70% removal efficiency. Interestingly, the growth of algal cells showed a growth trend of first inhibition and then promotion. After 96 h treatment, the SOD, CAT, MDA, and ROS levels of the phenol- and 4-fluorophenol-treated groups were similar or even lower than those of the control, proving that the growth of algal cells is stable after phenol or 4-fluorophenol exposure. Furthermore, it was revealed by a metabolomics study that significant upregulation of photosynthetic pigments and phospholipids could be caused by phenol and 4-fluorophenol treatment, benefitting the photosynthesis, antioxidation, and membrane stability of algal cells. Hence, it is indicated by these findings that *C. pyrenoidosa* is a promising biological material to remove such contaminants.

ACKNOWLEDGEMENTS

This study was funded by the Natural Science Foundation of Ningxia Province, P.R. China (grant no. 2022AAC05042), the Ningxia Key R&D projects (grant no. 2021BEG02006). The authors would like to express gratitude to EditSprings (https://www.editsprings.cn) for the expert linguistic services provided.

AUTHORS CONTRIBUTION

M.L. conceptualized the study, prepared methodology, acquired funds, wrote the original draft. L.M. investigated the study, did formal analysis, wrote the original draft. Y.A. led the investigation, did data analysis. D.W. wrote the review and edited the file, acquired funds. H.M. conceptualized and supervised the study. J.Z. and C.W. prepared the methodology and supervised the study.

DATA AVAILABILITY STATEMENT

All relevant data are included in the paper or its Supplementary Information.

CONFLICT OF INTEREST

The authors declare there is no conflict.

REFERENCES

Al-Khalid, T. & El-Naas, M. H. 2012 Aerobic biodegradation of phenols: a comprehensive review. *Critical Reviews in Environmental Science and Technology* **42**, 1631.

Andreotti, V., Solimeno, A., Rossi, S., Ficara, E., Marazzi, F., Mezzanotte, V. & García, J. 2020 Bioremediation of aquaculture wastewater with the microalgae *tetraselmis suecica*: semi-continuous experiments, simulation and photo-respirometric tests. *Science of The Total Environment* **738**, 139859.

Ansar, B. S. K., Kavusi, E., Dehghanian, Z., Pandey, J., Lajayer, B. A., Price, G. W. & Astatkie, T. 2022 Removal of organic and inorganic contaminants from the air, soil, and water by algae. *Environmental Science and Pollution Research*. https://doi.org/10.1007/s11356-022-21283-x.

Brycht, M., Lochyński, P., Barek, J., Skrzypek, S., Kuczewski, K. & Schwarzova-Peckova, K. 2016 Electrochemical study of 4-chloro-3-methylphenol on anodically pretreated boron-doped diamond electrode in the absence and presence of a cationic surfactant. *Journal of Electroanalytical Chemistry* **771**, 1–9.

Chen, S., Xie, J. & Wen, Z. 2021a Removal of pharmaceutical and personal care products (PPCPs) from waterbody using a revolving algal biofilm (RAB) reactor. *Journal of Hazardous Materials* **406**, 124284.

Chen, X., Wang, S., Sun, X. & Lu, Q. 2021b Cultivation of energy microalga *Chlorella vulgaris* with low–toxic sludge extract. *Water Science & Technology* **83**, 818–830.

Chen, Y.-D., Zhao, C., Zhu, X.-Y., Zhu, Y. & Tian, R.-N. 2021c Multiple inhibitory effects of succinic acid on *Microcystis aeruginosa*: morphology, metabolomics, and gene expression. *Environmental Technology* **43**, 3121–3130.

Chin, Y.-Y., Chu, W.-L., Kok, Y.-Y., Phang, S.-M., Wong, C.-Y., Tan, B.-K. & Mustafa, E. M. 2019 Sensitivity of selected tropical microalgae isolated from a farmland and a eutrophic lake to atrazine and endosulfan. *Journal of Applied Phycology* **31**, 2981–2998.

Couto, E., Assemany, P. P., Carneiro, G. C. A. & Soares, D. C. F. 2022 The potential of algae and aquatic macrophytes in the pharmaceutical and personal care products (PPCPs) environmental removal: a review. *Chemosphere* **302**, 134808.

Ding, T., Yang, M., Zhang, J., Yang, B., Lin, K., Li, J. & Gan, J. 2017 Toxicity, degradation and metabolic fate of ibuprofen on freshwater diatom *Navicula* sp. *Journal of Hazardous Materials* **330**, 127–134.

Duan, X., Sui, X., Wang, W., Bai, W. & Chang, L. 2019 Fabrication of a hydrophobic SDBS-PbO_2 anode for electrochemical degradation of nitrobenzene in aqueous solution. *Applied Surface Science* **494**, 211–222.

Escudero, A., Hunter, C., Roberts, J., Helwig, K. & Pahl, O. 2020 Pharmaceuticals removal and nutrient recovery from wastewaters by *Chlamydomonas acidophila*. *Biochemical Engineering Journal* **156**, 107517.

Fawzy, M. A. & Alharthic, S. 2021 Cellular responses and phenol bioremoval by green alga *Scenedesmus abundans*: equilibrium, kinetic and thermodynamic studies. *Environmental Technology & Innovation* **22**, 101463.

Ge, T., Han, J., Qi, Y., Gu, X., Ma, L., Zhang, C., Naeem, S. & Huang, D. 2017 The toxic effects of chlorophenols and associated mechanisms in fish. *Aquatic Toxicology* **184**, 78–93.

Hu, X., Ouyang, S., Mu, L., An, J. & Zhou, Q. 2015 Effects of graphene oxide and oxidized carbon nanotubes on the cellular division, microstructure, uptake, oxidative stress, and metabolic profiles. *Environmental Science & Technology* **49**, 10825–10833.

Huang, H. L. & Wang, B. G. 2004 Antioxidant capacity and lipophilic content of seaweeds collected from the Qingdao coastline. *Journal of Agricultural and Food Chemistry* **52**, 4993–4997.

Huang, H., Xiao, X., Ghadouani, A., Wu, J., Nie, Z., Peng, C., Xu, X. & Shi, J. S. 2015 Effects of natural flavonoids on photosynthetic activity and cell integrity in *Microcystis aeruginosa*. *Toxins* **7**, 66–80.

Ioannides, I. M., Fasoula, D. A., Robertson, K. R. & Rebeiz, C. A. 1994 An evolutionary study of chlorophyll biosynthetic heterogeneity in green plants. *Biochemical Systematics and Ecology* **22**, 211–220.

Jalilian, N., Najafpour, G. D. & Khajouei, M. 2020 Macro and micro algae in pollution control and biofuel production – a review. *ChemBioEng Reviews* **7**, 18–33.

Juksu, K., Zhao, J.-L., Liu, Y.-S., Yao, L., Sarin, C., Sreesai, S., Klomjek, P., Jiang, Y.-X. & Ying, G.-G. 2019 Occurrence, fate and risk assessment of biocides in wastewater treatment plants and aquatic environments in Thailand. *Science of The Total Environment* **690**, 1110–1119.

Li, L.-H., Hao, L.-C. & Hong, Y. 2022 Responses of bloom-forming *Microcystis aeruginosa* to polystyrene microplastics exposure: growth and photosynthesis. *Water Cycle* **3**, 133–142.

Li-Beisson, Y., Thelen, J. J., Fedosejevs, E. & Harwood, J. L. 2019 The lipid biochemistry of eukaryotic algae. *Progress in Lipid Research* **74**, 31–68.

Lima, L. M. & Barreiro, E. J. 2005 Bioisosterism: a useful strategy for molecular modification and drug design. *Current Medicinal Chemistry* **12**, 23–49.

Liu, X.-Y., Hong, Y. & Gu, W.-P. 2021 Influence of light quality on *Chlorella growth*, photosynthetic pigments and high-valued products accumulation in coastal saline-alkali leachate. *Journal of Water Reuse and Desalination* **11**, 301–311.

Novák, Z., Harangi, S., Baranyai, E., Gonda, S., B-Béres, V. & Bácsi, I. 2020 Effects of metal quantity and quality to the removal of zinc and copper by two common green microalgae (*Chlorophyceae*) species. *Phycological Research* **68**, 227–235.

Organisation for Economic Co-operation and Development (OECD) 2006 *OECD Guidelines for the Testing of Chemicals, Section 2: Effects on Biotic Systems. Test No. 201: Freshwater Alga and Cyanobacteria, Growth Inhibition Test*. OECD Publishing, Paris.

Ong, H. C., Tiong, Y. W., Goh, B. H. H., Gan, Y. Y., Mofijur, M., Fattah, I. M. R., Chong, C. T., Alam, M. A., Lee, H. V., Silitonga, A. S. & Mahlia, T. M. I. 2021 Recent advances in biodiesel production from agricultural products and microalgae using ionic liquids: opportunities and challenges. *Energy Conversion and Management* **228**, 113647.

Patel, N., Shahane, S., Bhunia, B., Mishra, U., Chaudhary, V. K. & Srivastav, A. L. 2022 Biodegradation of 4-chlorophenol in batch and continuous packed bed reactor by isolated *Bacillus subtilis*. *Journal of Environmental Management* **301**, 113851.

Pinckney, J. L., Hagenbuch, I. M., Long, R. A. & Lovell, C. R. 2013 Sublethal effects of the antibiotic tylosin on estuarine benthic microalgal communities. *Marine Pollution Bulletin* **68**, 8–12.

Qian, L., Qi, S., Cao, F., Zhang, J., Zhao, F., Li, C. & Wang, C. 2018 Toxic effects of boscalid on the growth, photosynthesis, antioxidant system and metabolism of *Chlorella vulgaris*. *Environmental Pollution* **242**, 171–181.

Rebeiz, C. A., Parham, R., Fasoula, D. A. & Ioannides, I. M. 1994 Chlorophyll a biosynthetic heterogeneity. *Ciba Foundation Symposium* **180**, 177–189.

Reddy, K., Renuka, N., Kumari, S. & Bux, F. 2021 Algae-mediated processes for the treatment of antiretroviral drugs in wastewater: prospects and challenges. *Chemosphere* **280**, 130674.

Sawaya, C., Khoury, C. E., Ramadan, L., Deeb, R. & Harb, M. 2022 Effects of influent municipal wastewater microbial community and antibiotic resistance gene profiles on anaerobic membrane bioreactor effluent. *Journal of Water Reuse and Desalination* **12**, 304–318.

Sutherland, D. L., Howard-Williams, C., Turnbull, M. H., Broady, P. A. & Craggs, R. J. 2015 Enhancing microalgal photosynthesis and productivity in wastewater treatment high rate algal ponds for biofuel production. *Bioresource Technology* **184**, 222–229.

Upchurch, R. G. 2008 Fatty acid unsaturation, mobilization, and regulation in the response of plants to stress. *Biotechnology Letters* **30**, 967–977.

Wang, X. C. 2022 Safe water reuse through a quasi-natural water cycle. *Journal of Water Reuse and Desalination* **12**, 366–372.

Xiao, G., Chen, J., Show, P. L., Yang, Q., Ke, J., Zhao, Q., Guo, R. & Liu, Y. 2021 Evaluating the application of antibiotic treatment using algae-algae/activated sludge system. *Chemosphere* **282**, 130966.

Xiong, J.-Q., Kurade, M. B., Abou-Shanab, R. A. I., Min-KyuJi, C., Kim, J. & and Byong-HunJeon, J. O. 2016 Biodegradation of carbamazepine using freshwater microalgae *Chlamydomonas mexicana* and *Scenedesmus obliquus* and the determination of its metabolic fate. *Bioresource Technology* **205**, 183–190.

Yang, Y., Xu, J., Huang, L., Leng, Y., Dai, L., Rao, Y., Chen, L., Wang, Y., Tu, Z., Hu, J., Ren, D., Zhang, G., Zhu, L., Guo, L., Qian, Q. & Zeng, D. 2016 PGL, encoding chlorophyllide a oxygenase 1, impacts leaf senescence and indirectly affects grain yield and quality in rice. *Journal of Experimental Botany* **5**, 1297–1310.

Zhai, Q., Hong, Y., Wang, X., Wang, Q., Zhao, G., Liu, X. & Zhang, H. 2022 Mixing starch wastewaters to balance nutrients for improving nutrient removal, microalgae growth and accumulation of high value-added products. *Water Cycle* **3**, 151–159.

Zhang, M., Steinman, A. D., Xue, Q., Zhao, Y., Xu, Y. & Xie, L. 2020 Effects of erythromycin and sulfamethoxazole on *Microcystis aeruginosa*: cytotoxic endpoints, production and release of microcystin-LR. *Journal of Hazardous Materials* **339**, 123021.

Zhu, Y., Xu, G., Wang, X., Ji, X., Jia, X., Sun, L., Gu, X. & Xie, X. 2022 Passive sampling of chlorophenols in water and soils using diffusive gradients in thin films based on *β*-cyclodextrin polymers. *Science of The Total Environment* **806**, 150739.

First received 24 December 2022; accepted in revised form 11 February 2023. Available online 10 March 2023

doi: 10.2166/wrd.2023.084

Removal of antibiotics with different charges in water by graphene oxide membranes

Zimeng Liang, Xin Zhao *, Weiqi Huang, Huabiao Qi and Can Wang

School of Environmental Science and Engineering, Tianjin University, Tianjin 300350, China
*Corresponding author. E-mail: xinzhao@tju.edu.cn

XZ, 0000-0001-6042-5762

ABSTRACT

Antibiotics are a large group of emerging organic pollutants with low concentration levels in the water. The presence of antibiotics will affect the ecological environment and human health. The removal of trace organic compounds by graphene oxide (GO) membranes has attracted extensive attention. This study investigated the removal of three differently charged antibiotics by GO membranes and the influence of water quality on the removal of antibiotics. It showed that a crosslinked ethylenediamine-GO (EDA-GO) membrane had better stability and higher antibiotic removal performance than a non-crosslinked GO membrane. Among the three antibiotics, penicillin (PNC) was negatively charged and had the highest removal efficiency due to steric effect and electrostatic repulsion. A low concentration (10 mmol L^{-1}) of Na^+ in water could increase the membrane flux but had no significant effect on the removal of antibiotics. Ca^{2+} could reduce the membrane flux and improve the removal of chloramphenicol (CAP) and PNC. The removal efficiencies of low-concentration antibiotics (500 μg L^{-1}) were higher than those of high-concentration antibiotics (10 mg L^{-1}). Furthermore, the removal of antibiotics under the condition of actual wastewater quality was higher than those in solutions prepared with ultrapure water. The EDA-GO membrane has great potential in the removal of antibiotics in wastewater.

Key words: antibiotic, coexisting cation, crosslinked GO membrane, nanofiltration, wastewater reclamation and reuse

HIGHLIGHTS

- The removal of three differently charged antibiotics by GO membranes was studied.
- Crosslinked EDA-GO membrane had higher removal performance and better stability.
- The removal of PNC with a negative charge was higher than CAP and ERY.
- Ca^{2+} could reduce the membrane flux and increase the removal of antibiotics.
- The removal of antibiotics under the condition of actual wastewater quality was higher.

GRAPHICAL ABSTRACT

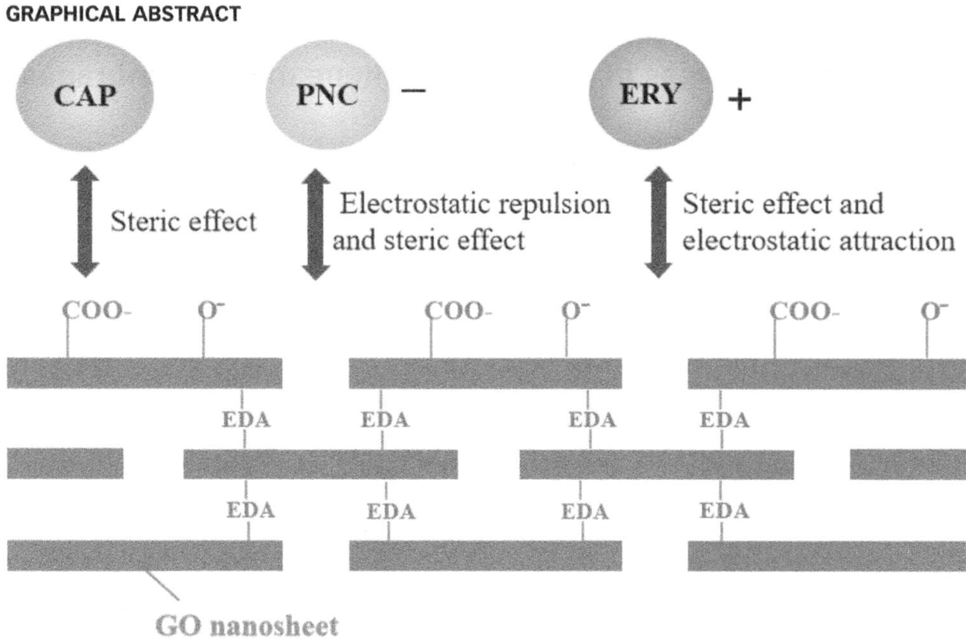

1. INTRODUCTION

As a kind of emerging pollutant, antibiotics in the aquatic environment and their potentially harmful effects on aquatic organisms have aroused widespread concern around the world (Liu *et al.* 2020). Antibiotics in water may seriously interfere with various physiological functions of the human body, damage the human immune system, bring adverse effects to the aquatic ecosystem, and promote the production of antibiotic resistance of microorganisms in the environment (Wallmann *et al.* 2021). Therefore, the removal of such pollutants is particularly important. In recent years, nanofiltration technology has been widely welcomed because it can remove trace organics in drinking water or reclaimed water (Nghiem *et al.* 2004; Kong *et al.* 2016). However, the traditional nanofiltration membrane has the disadvantages of poor anti-pollution and oxidation resistance. So, it still cannot operate well in treating wastewater with high concentrations of pollutants/salts under adverse conditions (Wang *et al.* 2020; Yuan *et al.* 2020).

Graphene oxide (GO) membrane has attracted extensive attention due to its inherent physical and chemical properties, such as adjustable surface function, excellent oxidation resistance, and high thermal stability (Hu & Mi 2013; Dong *et al.* 2016; Ying *et al.* 2016; Zhao *et al.* 2016).

GO is generated from graphene through a series of oxidation reactions. Since there are rich oxygen-containing functional groups (such as –COOH, –OH, etc.) on the edge and base surface, GO nanosheets can be easily modified and adjusted to adapt to different applications (Yu *et al.* 2015). In addition, these oxygen-containing functional groups enable GO nanosheets to have excellent dispersibility in water, which can further improve the processability of GO nanosheets as membrane materials through different methods (such as vacuum suction filtration, spin coating, and layer-by-layer assembly) (Liang *et al.* 2016; Nam *et al.* 2016; Karunakaran *et al.* 2017). Due to its compact layered structure and controllable layer spacing (the distance between adjacent GO nanosheets), nanofiltration membranes based on GO materials have proved to have excellent performance in the removal of organics in water (Hung *et al.* 2014; Lecaros *et al.* 2017).

Similar to the interception mechanism of the polymer nanofiltration membrane, GO-based membranes can become a selective separation membrane through various effects (steric effect, solute GO interaction, electrostatic effect, and adsorption). Previous studies have shown that the removal efficiencies of different antibiotics by the GO membrane vary greatly, and the retention mechanism mainly depends on membrane material characteristics and the physical and chemical properties of target substances (Bellona *et al.* 2004; Khanzada *et al.* 2020; Kong *et al.* 2014, 2015). Many research works showed that the GO membrane had high retention performance for small molecule dyes (such as new carmine, methyl blue, and Congo red) (Zhang *et al.* 2017a, 2017b; Song *et al.* 2018; Qi *et al.* 2020). The good adsorption mechanism of GO enables

GO membranes to have high adsorption performance for antibiotic molecules. Since the surface of the GO membrane has negative charges, the removal mechanisms of the membrane for antibiotics with different charges in water may be different. If the performance and mechanism of the GO membrane in antibiotics interception can be fully understood, it will be very helpful for nanofiltration technology to remove antibiotics in the future. Therefore, it is necessary to verify the applicability of the GO membrane in the interception of antibiotics.

In this study, typical antibiotics with different charges in wastewater were selected as the research object. The GO membrane was prepared by the vacuum filtration method, and its removal effect on antibiotics was investigated. Furthermore, the influences of membrane material, coexisting cation, pH, and antibiotic concentration on antibiotics removal performance were analyzed.

2. MATERIALS AND METHODS

2.1. Materials and chemicals

Commercial GO powder was purchased from XFNANO (XF002-2, Nanjing XFNANO Materials Tech Co., Ltd, China). Characterized by the manufacturer, GO powder had a film diameter of 0.5–5 μm and a thickness of 0.8–1.2 nm. Polyvinylidene fluoride (PVDF) membranes with a diameter of 63.5 mm and a nominal pore size of 0.22 μm (GVWP06225, Millipore, Billerica, MA) were used as the support layer for the GO membrane fabrication. Three substances represented by antibiotics were purchased from TCI (Chemical Industry Development Co., Ltd, Shanghai). In solution, they are uncharged chloramphenicol (CAP), penicillin (PNC) with a negative charge, and erythromycin (ERY) with a positive charge, and their chemical structures are shown in Supplementary material, Figure S1. Inorganic salts including NaCl, KCl, $MgCl_2$, $CaCl_2$, and KOH were of analytical grade. All solutions were prepared using ultrapure (UP) water (Q-Gard$^®$ 1, Millipore, France).

2.2. Fabrication of GO membranes

All GO membranes in this study were fabricated by vacuum filtration. For the non-crosslinked membranes, GO suspensions ($0.005\ mg^{-1}\ mL^{-1}$) were prepared by sonication for 2 h. Then, 52 mL of the GO suspensions were filtered through the PVDF membrane by vacuum filtration under 0.8 bar at room temperature. The obtained membrane was dried at 40 °C for 40 h to remove water molecules trapped within the GO layers.

For the crosslinked membranes, 5 mL of ethylenediamine (EDA, 99.5 wt%) was added dropwise to 500 mL of GO suspension ($0.005\ mg\ mL^{-1}$). The mixed solution was stirred for 3 h (temperature 30 °C, rotation speed 200 rpm) and then cooled to room temperature. After that, the filtration process was the same as that of the non-crosslinked membrane described previously, and finally, an EDA covalently crosslinked GO membrane (EDA-GO membrane) is obtained (Supplementary material, Figure S2).

2.3. Characterization of GO membranes

The membrane morphology was characterized by a scanning electron microscope (SEM, HitachiS4800, Japan), X-ray photoelectron spectroscopy (XPS, PHI5000VersaProbe, ULVCA-PHI, USA), a Fourier transform infrared (FTIR) spectrometer (Vertex 70, Bruker, Karlsruhe, Germany), and X-ray diffraction (XRD, Bruker D8 Advance, Karlsruhe, Germany). The d-spacing of the non-crosslinked GO membrane and the EDA-GO membrane under different conditions was calculated by XRD. Scanning was conducted at a speed of 3°/min, for 2θ from 5° to 25°.

The stability of the GO membrane was investigated by placing different GO membranes in UP water, recording the appearance changes of the inner membrane from 0 to 40 days, and observing the shedding of the GO functional layer from the PVDF support layer.

2.4. GO membrane filtration tests

The GO membrane fluxes were tested using a dead-end filtration system at room temperature. In this process, the feed solution was loaded onto a stirred cell (Merck Millipore, Germany) with an effective membrane area of 13.4 cm^2 from a feed tank of 800 mL (Merck Millipore, Germany). Before the filtration of the solutions, UP water was fed into the cell under 4 bar to compact the GO membrane for about 2 h until the water flux was stable. The feed was then replaced by an antibiotic solution. The filtration was conducted under the same pressure (4 bar) for 5 h. Changes in effluent flux were recorded every 5 min by a computer connected to the electronic balance. Permeate samples were collected every 1 h to evaluate the concentration of antibiotics in the effluent. The antibiotic concentrations of the feed used in this study were 500 µg L^{-1} and 10 mg L^{-1}.

2.5. Analytical methods of antibiotics concentration

Antibiotic samples with lower concentrations ($<$1 mg $\cdot$ L^{-1}) were detected by ultrahigh performance liquid chromatography-mass spectrometry (UPLC-MS); 0.1% formic acid aqueous solution and acetonitrile were used as the mobile phase, the flow rate of the mobile phase was 0.4 mL$\cdot$min^{-1}, the chromatographic column temperature was 40 °C, and the injection volume was 10 µL (Supplementary material, Table S1).

Antibiotic samples with higher concentrations (1–10 mg$\cdot$L^{-1}) were determined by the high-performance liquid chromatography (HPLC) method. The CAP solution was detected by an ultrahigh-performance liquid phase ultraviolet detector. Acetonitrile and 0.3% formic acid aqueous solution were used as the mobile phase, the constant ratio was 2:1, the flow rate was 0.1 mL$\cdot$min^{-1}, the chromatographic column temperature was 35 °C, and the injection volume was 10 µL. The wavelength of the UV detector was 280 nm, and the running time of each sample was 4 min. PNC solution was also detected by an ultrahigh-performance liquid phase ultraviolet detector. Methanol and 0.001 mol$\cdot$L^{-1} of potassium dihydrogen phosphate solution was used as the mobile phase, the constant ratio was 60:40, the flow rate was 0.15 mL$\cdot$min^{-1}, the temperature of the chromatographic column was 35 °C, and the injection volume was 10 µL. The wavelength of the UV detector was 225 nm, and the running time of each sample was 4.5 min. The ERY solution after 20 times dilution was detected by the UPLC-MS method which was used for low-concentration solutions.

2.6. Actual wastewater sample

In this study, we collected the effluents of secondary biological treatment and ultrafiltration processes from a wastewater treatment plant to study the removal of antibiotics under the condition of actual wastewater quality by the EDA-GO membrane. Antibiotics were added into actual wastewater samples, in which the concentration of these three antibiotics was lower than the detection limit (1–10 µg L^{-1}), and the samples containing 10 mg$\cdot$L^{-1} of CAP, PNC, and ERY were prepared, respectively. The water quality of the two samples collected is shown in Table 1.

2.7. Statistical analysis

All data were analyzed using SPSS 27.0 software. The significant differences were evaluated by ANOVA. The *P* values of $<$0.05 were considered significant and expressed as '*'.

3. RESULTS AND DISCUSSION

3.1. Characterization of GO membranes

SEM images showed that the surface morphology of the non-crosslinked GO membrane and the EDA-GO membrane show a certain degree of wrinkles, the wrinkles of the EDA-GO membrane are more uniform, and there are no obvious defects in both membranes (Supplementary material, Figure S3). XPS results showed that the GO membrane contains C–O, C–C and C=C bonds. The strength of C–O and C–C bonds in the EDA-GO membrane decreased and C–N bonds appeared, which further indicated the crosslinking reaction between EDA and GO during the fabrication of the EDA-GO membrane (Supplementary material, Figure S4). In the FTIR spectra, the C=O absorption peak at 1,740 cm^{-1}, the C–O tensile vibration peak in the epoxy group near 1,620 cm^{-1}, and the –OH absorption peak at 1,000 cm^{-1}, indicated that there were many hydrophilic oxygen-containing functional groups on the surfaces of the two GO membranes. Also, the stretching vibration peak of the O=C–NH group of the EDA-GO membrane at 1,574 cm^{-1} indicated that the COOH of GO interacted with NH$_2$ of EDA and generated the O=C–NH group (Supplementary material, Figure S6) (Meng *et al.* 2018; Kong *et al.* 2020).

Moreover, both dry and wet (immersed in UP water) EDA-GO membranes were characterized by XRD. Figure 1 shows that the d-spacing of the EDA-GO membrane immersed in solution is larger than that of the dry membrane.

The prepared non-crosslinked GO membrane and the EDA-GO membrane were immersed in UP water, and the stability of the two membranes was judged by investigating the peeling of the GO functional layer from the PVDF support layer. The results showed that the non-crosslinked GO membrane had an obvious swelling effect when immersed in water, and that the stable chemical bond formed in the EDA-GO membrane can reduce the swelling effect. As a result, the EDA-GO membrane is much more stable than the non-crosslinked GO membrane (Supplementary material, Figure S7).

Table 1 | The water quality of wastewater samples

Index	Secondary effluent	Ultrafiltration effluent
pH	7.1	7.1
TOC (mg·L^{-1})	36.2	3.6
COD (mg·L^{-1})	22	<1
TN (mg·L^{-1})	4.6	4.6
NH$_3$-N (mg·L^{-1})	<1	<1
TP (mg·L^{-1})	0.77	0.59
Conductivity (µS·cm^{-1})	1,429	1,380

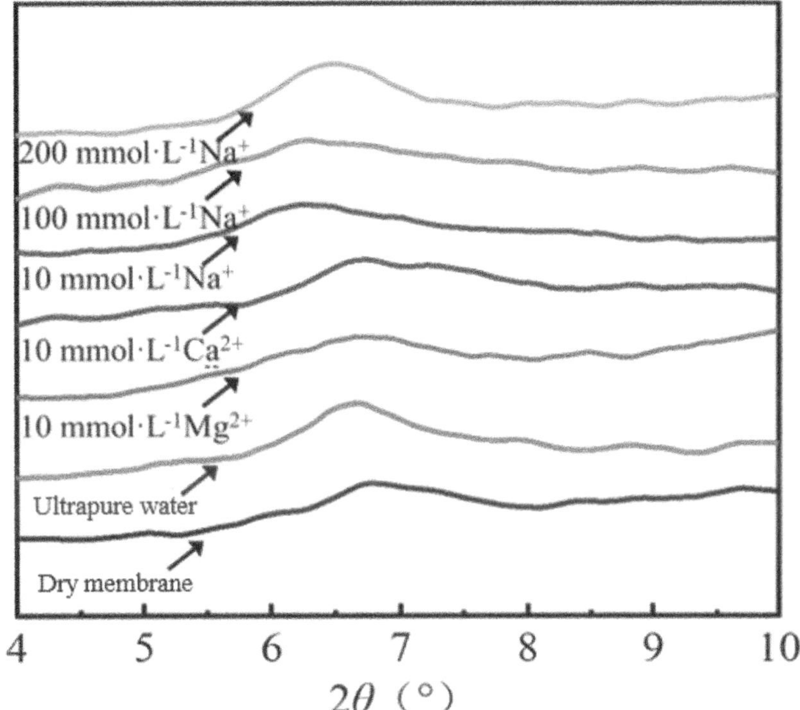

Figure 1 | XRD diffraction patterns of EDA-GO membranes under different conditions.

3.2. Removal of antibiotics by the GO membrane

3.2.1. Comparison of antibiotics removal by different GO membranes

The EDA-GO membrane was used to filter three antibiotics solutions of 10 mg·L^{-1} CAP, PNC, and ERY, respectively. The pure water flux J_{UP} of the non-crosslinked GO membrane and the EDA-GO membrane was 12.6 and 7.5 L·m^{-2}·h^{-1}, respectively (Figure 2(a)).

Figure 2(b) shows that the average removal efficiencies of CAP, PNC, and ERY solutions by the non-crosslinked GO membrane were 52, 84, and 72%, respectively, while those were 57, 92, and 81% by the EDA-GO membrane. Due to better stability and smaller d-spacing of the EDA-GO membrane, the removal of PNC and ERY was higher than that of the non-crosslinked membrane, while the removal of CAP had no significant difference.

The removal mechanism of the GO membrane for organic substances with different charges is different. For electrically neutral CAP, the removal mechanism was mainly the steric effect. The removal mechanism of negatively charged PNC was mainly electrostatic repulsion and the steric effect. The positively charged ERY had a large molecule, and the steric hindrance effect was obvious. In addition, ERY molecules entering the membrane nano channel have an electrostatic attraction with the negatively charged GO nanosheets and stay inside the membrane.

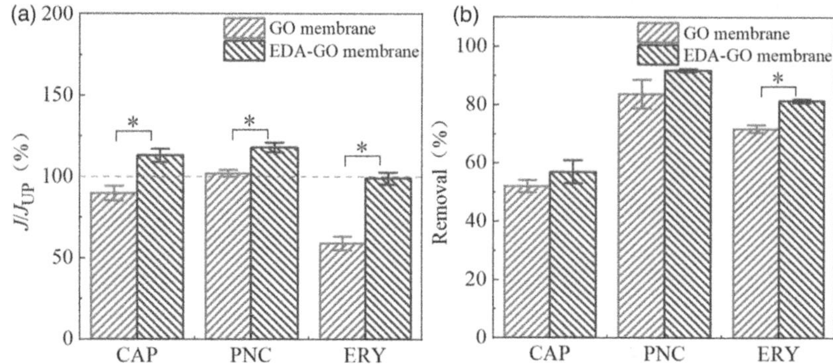

Figure 2 | (a) $J//J_{UP}$ of antibiotic solutions filtered by different GO membranes and (b) removal of antibiotics by different GO membranes.

3.2.2. Removal of antibiotics by the EDA-GO membrane with different loads

The performance of the nanofiltration membrane was related not only to the material of the membrane itself but also to the load of the membrane material. In this study, the EDA-GO membranes with a load of 50, 100, and 150 mg·m^{-2} were used to filter CAP, PNC, and ERY solutions with a concentration of 10 mg·L^{-1}.

By comparing $J//J_{UP}$ of the three antibiotic solutions filtered by EDA-GO membranes with different loads, it showed that the average $J//J_{UP}$ of the three antibiotic solutions filtered by the membrane with a load of 150 mg·m^{-2} was the highest (Figure 3(a)).

The removal efficiencies of the three antibiotics by the EDA-GO membrane with a load of 50 mg·m^{-2} were the lowest (Figure 3(b)). Meanwhile, the removal increased with the increase in load, but it did not change significantly when the load was greater than 100 mg·m^{-2}.

When the load of the membrane was too low, there had been uneven coverage or damage on the membrane surface during the membrane fabrication process, and the GO layer was too thin to intercept antibiotic molecules. As the membrane thickness increased, the ability of the EDA-GO membrane to intercept antibiotic molecules was enhanced. When the load was greater than 100 mg·m^{-2}, membrane thickness was no longer the influencing factor of antibiotic removal.

3.2.3. Mechanism of antibiotic removal by the EDA-GO membrane

The adsorption performance of the EDA-GO membrane on antibiotics was investigated, so as to further explore the removal mechanism of antibiotics. The EDA-GO membranes were put into solutions of 10 mg·L^{-1} CAP, PNC, and ERY for 24 h, respectively. So, the changes in antibiotic concentrations were measured to explore the adsorption performance of the EDA-GO membrane for the three antibiotics.

The results showed that the adsorption capacity of the EDA-GO membrane for the three antibiotics was extremely low, less than 0.02 mg·L^{-1} after 24 h, which was less than 0.2% of the initial concentration (Figure 4). The removal mechanism of the

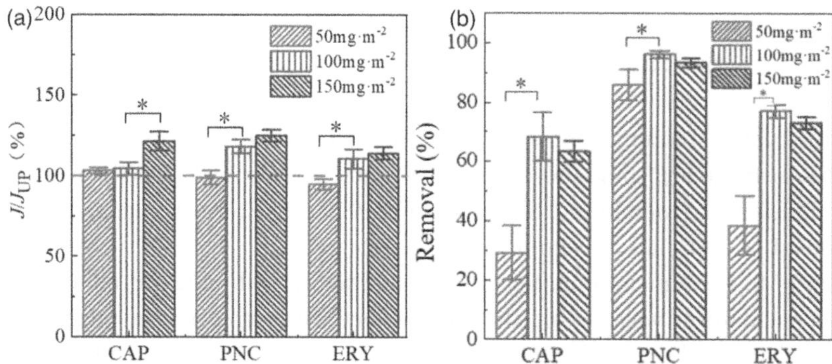

Figure 3 | (a) $J//J_{UP}$ of antibiotic solutions filtered by the EDA-GO membrane with different loads and (b) removal of antibiotics by the EDA-GO membrane with different loads.

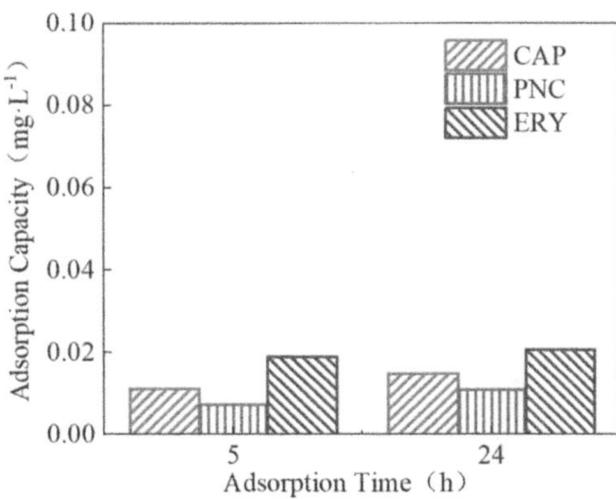

Figure 4 | Adsorption of antibiotics by the EDA-GO membrane.

nanofiltration membrane for trace organics was mainly steric effect, electrostatic effect, and adsorption (Bellona *et al.* 2004; Verliefde *et al.* 2009; Steinle-Darling *et al.* 2010). Since the adsorption of the three antibiotics by the EDA-GO membrane was not obvious, it can be inferred that its retention mechanism was mainly the steric effect and the electrostatic effect (Zhao *et al.* 2019).

The removal of antibiotics mainly depended on the characteristics of the membrane and its physical and chemical properties (Kong *et al.* 2016; Mahlangu *et al.* 2016; Khanzada *et al.* 2020). CAP was electrically neutral in water, thus the mechanism of removing CAP by the EDA-GO membrane was mainly the steric effect (Han *et al.* 2016; Hao *et al.* 2017). PNC was negatively charged in water, while the EDA-GO membrane contained rich oxygen-containing functional groups (such as –COOH, –OH, epoxy group, etc.) with a negative charge (Chu *et al.* 2017; Zhao *et al.* 2019). Therefore, the mechanism of the EDA-GO membrane to remove PNC mainly included the steric effect and electrostatic repulsion. ERY was positively charged in water. When ERY molecules entered the inner nanochannel of the EDA-GO membrane, they had an electrostatic attraction with GO nanosheets with a negative charge, resulting in a part of ERY molecules staying inside the EDA-GO membrane (Xu *et al.* 2018). Therefore, the mechanism of the EDA-GO membrane to remove ERY mainly included the steric effect and the electrostatic effect. In addition, the size of the ERY molecule was larger than the other two antibiotics, and its steric effect was more significant.

3.3. The influence of water quality on the removal of antibiotics by the EDA-GO membrane

3.3.1. Effect of coexisting cations

Inorganic ions in actual wastewater can adjust the d-spacing between independent GO layers during the filtration process, thus changing the structure of the EDA-GO membrane and its separation performance (Kang *et al.* 2018). In this study, the changes of flux and removal of antibiotics during filtration were investigated in the presence of coexisting cations (10, 100, and 200 mmol·L^{-1} Na$^+$, 10 mmol·L^{-1} Mg^{2+}, and 10 mmol·L^{-1} Ca^{2+}, respectively).

When the EDA-GO membrane was used to filter antibiotic solutions with coexisting Na$^+$, $J//J_{UP}$ was relatively higher. When filtering the solution with the coexisting ions of Ca^{2+}, $J//J_{UP}$ was relatively lower (Figure 5(a)).

The previous study showed that Na$^+$ could cause the increase of d-spacing of the GO membrane, which reduced the van der Waals force between GO nanosheets, and the electrostatic attraction between Na$^+$ and GO nanosheets was too weak to form a bridge between them to compensate for the decrease of van der Waals force, so the flux increased when the solution contained Na$^+$ (Xu *et al.* 2018). When Ca^{2+} was present, a strong complexation reaction occurred between Ca^{2+} and GO. Ca^{2+} was closely combined with –COOH at the edge of the GO nanosheet and acted as a crosslinking agent, which crosslinked the single GO nanosheets together, increased the stability of the membrane, and led to the reduction of d-spacing of the GO layer. As a result, $J//J_{UP}$ was reduced.

For the removal of antibiotics, when CAP and ERY solutions were filtered, the highest removal efficiencies were obtained in the presence of Ca^{2+} (Figure 5(b)). Meanwhile, the removal of PNC did not change significantly with the coexisting cations and was higher than those of CAP and ERY.

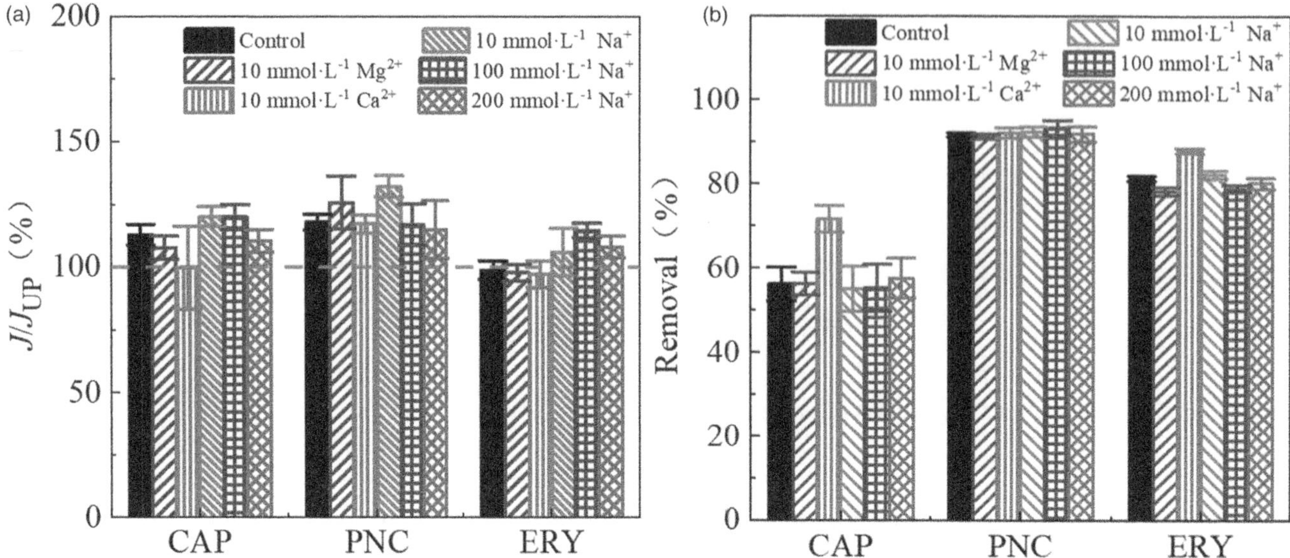

Figure 5 | (a) Comparison of $J//J_{UP}$ in the filtration of antibiotic solution with different coexisting cations by the EDA-GO membrane and (b) removal of antibiotics by the EDA-GO membrane with different coexisting cations.

When filtering CAP and ERY solutions, Ca^{2+} had a strong crosslinking effect with –COOH in GO, resulting in a decrease in d-spacing, thus increasing the removal of antibiotics due to the steric effect. For PNC, the removal was generally higher than 90% by the EDA-GO membrane, so the influence of Ca^{2+} was not as obvious as that for CAP and ERY.

Previous studies have shown that inorganic ions can significantly change the d-spacing of the non-crosslinked GO membrane, increase the effluent flux, and reduce the removal of target pollutants (Mi 2014; Mo *et al.* 2016). This study showed that coexisting cations had a small impact on membrane flux and removal performance. It demonstrated that the stability of the EDA-GO membrane was higher.

3.3.2. Effect of pH

pH can affect the structure of the EDA-GO membrane, thus affecting its pollutant removal performance (Huang *et al.* 2013; Yeh *et al.* 2015). This study investigated the effect of pH (6–9) on the flux and antibiotic removal performance of the EDA-GO membrane.

As shown in Figure 6(a), with the increase of pH, $J//J_{UP}$ of CAP and PNC solutions slightly increased, while that of the ERY solution slightly decreased. But they did not change significantly. This is because the dissociation reaction of functional

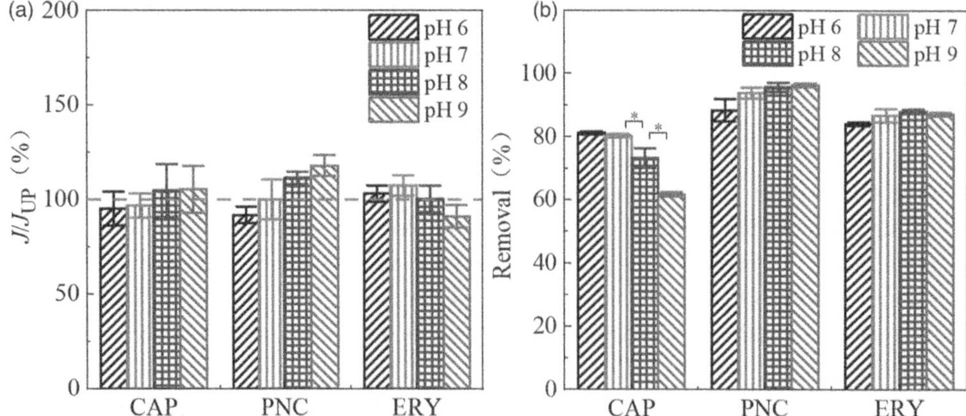

Figure 6 | (a) Comparison of $J//J_{UP}$ in the filtration of antibiotic solutions by the EDA-GO membrane at different pH conditions and (b) removal of antibiotics by the EDA-GO membrane at different pH conditions.

groups in GO in the solution moves to the right with the increase of pH (Supplementary material, Figure S8), and more negatively charged ions are generated (Choi *et al.* 2006; Ganesh *et al.* 2013). The increase of negative charges increases the electrostatic repulsion force between the GO layers, increases d-spacing, and reduces the hydrodynamic resistance of water molecules through the membrane, resulting in the increase of J/J_{UP}. In addition, for the PNC solution, the larger the pH was, the greater the electrostatic repulsive force between negatively charged PNC molecules and GO, which increased the d-spacing. However, the ERY molecule had a positive charge, and the ERY molecule entering the membrane had an electrostatic attraction with the GO nanosheet and reduced the d-spacing, leading to a slight decline in J/J_{UP}.

The removal of the three antibiotics by the EDA-GO membrane under different pH conditions was compared. The results showed that with the increase of pH, the removal of CAP gradually decreased, the removal of PNC gradually increased, while the removal of ERY did not change significantly (Figure 6(b)). When filtering the CAP solution, the increase of pH led to the increase of d-spacing of the EDA-GO membrane. As CAP was electrically neutral in water, the main interception mechanism for removing CAP was the steric effect. The larger the d-spacing was, the weaker the steric effect, resulting in the lower removal efficiency of CAP. PNC molecules were negatively charged in water, and the electrostatic repulsion between GO and PNC was the main removal mechanism. The increase in pH led to more negative charges on the GO membrane, and a stronger electrostatic repulsion between GO and PNC increased the removal of PNC. ERY molecules were positively charged. On the one hand, an increase in pH led to an increase in the d-spacing of the EDA-GO membrane. On the other hand, the electrostatic attraction between GO and ERY molecules increased, resulting in a decrease in the d-spacing of the membrane. The effects of the two actions on the removal performance were opposite, so the removal of ERY had no obvious change.

3.3.3. Effect of organic composition and concentration

The interaction between different organic substances may have an impact on their removal during the membrane filtration process. A mixed solution of CAP, PNC, and ERY was prepared and was filtered by the EDA-GO membrane. The results were compared with the single antibiotic solution. The results showed that the average J/J_{UP} when filtering the mixed solution was 104%, while the average J/J_{UP} of the single antibiotic solution was 113, 118, and 99% for CAP, PNC, and ERY, respectively (Supplementary material, Figure S9(a)). Also, the removal efficiencies of the three antibiotics in the mixed solution were similar to those in the single antibiotic solution (Supplementary material, Figure S9(b)). It indicated that the interaction between CAP, PNC, and ERY did not affect their rejection performances by the EDA-GO membrane.

The concentration of antibiotics in actual municipal wastewater is relatively low, generally in the range of $ng·L^{-1}$–$\mu g·L^{-1}$. This study investigated the impact of antibiotic concentration on its removal by the EDA-GO membrane. The results showed that the average J/J_{UP} of low-concentration antibiotic solutions (500 $\mu g·L^{-1}$) was 114, 126, and 113% for CAP, PNC, and ERY, respectively, while those of high-concentration antibiotic solutions (10 $mg·L^{-1}$) were 113, 118 and 99% (Figure 7(a)). Generally, the membrane flux decreased slightly with the increase in antibiotic concentration.

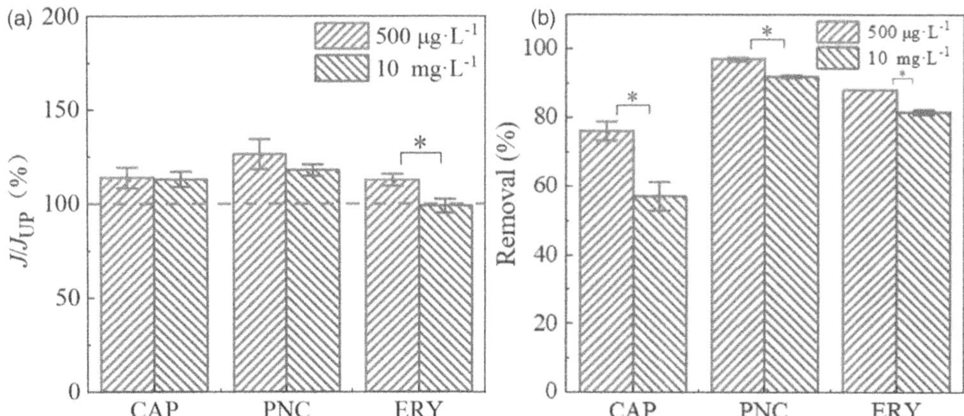

Figure 7 | (a) J/J_{UP} in the filtration of antibiotic solutions of different concentrations by the EDA-GO membrane and (b) removal of antibiotics of different concentrations by the EDA-GO membrane.

For the removal of antibiotics by the EDA-GO membrane, the results showed that the removal efficiencies of antibiotics in low-concentration solutions were 76, 97, and 88% for CAP, PNC, and ERY, respectively (Figure 7(b)). When the antibiotic concentration was 10 mg·L^{-1}, the removal efficiencies of CAP, PNC, and ERY by the EDA-GO membrane were 57, 92, and 81%, respectively. In general, the removal efficiencies of the three antibiotics with lower concentrations were higher than those with higher concentrations.

3.3.4. Removal of antibiotics under the condition of actual wastewater quality

The composition of organic and inorganic substances in actual wastewater is an extremely complex system, where some pollutants affect the separation performance of the EDA-GO membrane. This study investigated the removal of antibiotics under the condition of wastewater quality by adding antibiotics to actual wastewater samples.

The results showed that $J//J_{UP}$ of the secondary effluent filtered by the EDA-GO membrane was about 78%, and $J//J_{UP}$ of the ultrafiltration effluent was 75–85% (Figure 8(a)). When using the EDA-GO membrane to filter antibiotic solutions prepared by actual secondary effluent and ultrafiltration effluent samples, $J//J_{UP}$ was much lower than those of antibiotic solutions prepared by UP water. As the pollutants in actual wastewater are more complex than those in antibiotic solutions prepared by UP water, the pollutants narrow nanochannels of the EDA-GO membrane or even block the membrane during filtration, resulting in increased hydraulic resistance of water molecules through the membrane and reduced membrane flux. Therefore, $J//J_{UP}$ of secondary and ultrafiltration effluents was far lower than those of antibiotic solutions prepared by UP water. Since the concentration of organic matters (TOC and COD values in Table 1) in the ultrafiltration effluent was lower than that in the secondary effluent, the membrane flux when filtering the ultrafiltration effluent was slightly higher than that when filtering the secondary effluent.

The removal of antibiotics by the EDA-GO membrane was compared under different water quality conditions (Figure 8(b)). It showed that the removal efficiencies of antibiotics by the EDA-GO membrane in the solutions prepared by UP water were 57–90%, while the removal efficiencies of antibiotics in secondary and ultrafiltration effluents were 85–98%. The removal efficiencies of antibiotics in secondary and ultrafiltration effluents were higher than those in the solutions prepared by UP water. This was because the blocking effect of organic matters in wastewater samples increased the steric resistance effect of the EDA-GO membrane to intercept antibiotics (Xu *et al.* 2018). The interception mechanism of the EDA-GO membrane to intercept CAP was mainly the steric effect. The removal of CAP in secondary and ultrafiltration effluents was much higher than that in the solution prepared by UP water. Meanwhile, the interception mechanism of the EDA-GO membrane to PNC and ERY included the electrostatic effect and the steric effect, so the removal performances of the EDA-GO membrane to PNC and ERY were not quite different under three different water quality conditions.

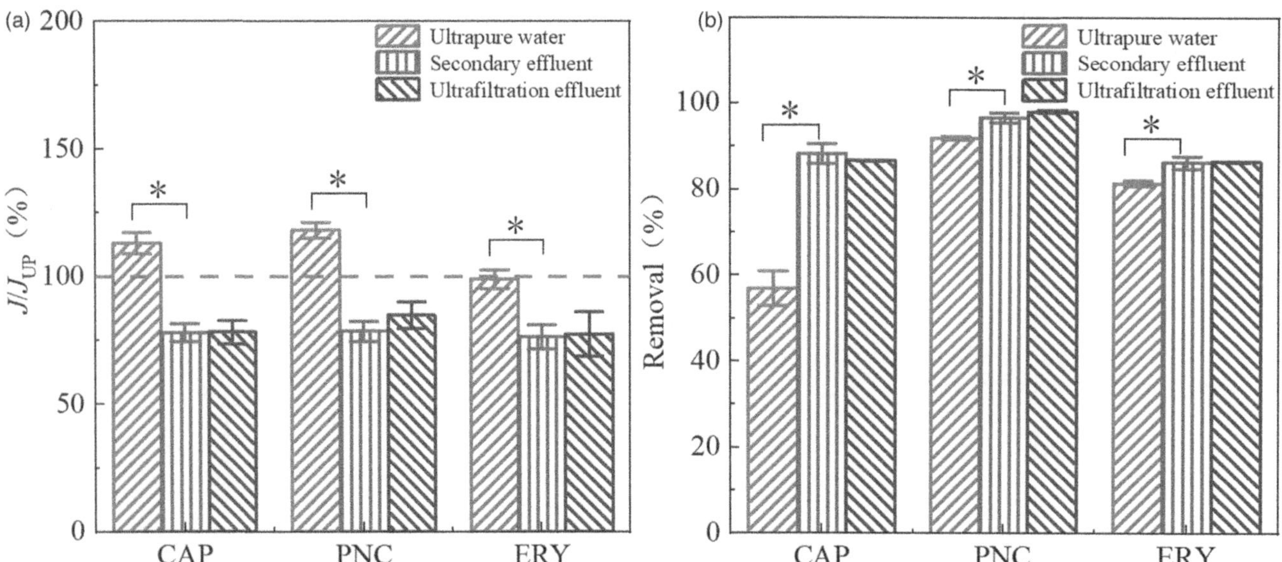

Figure 8 | (a) $J//J_{UP}$ of antibiotic solutions filtered by the EDA-GO membrane under different water quality conditions and (b) removal of antibiotics in UP water solution and actual wastewater filtered by the EDA-GO membrane.

4. CONCLUSIONS

This study investigated the removal performance of antibiotics with different charges in water by the GO membrane, explored the influence of water quality conditions (coexisting inorganic ions, pH, organic matter concentration, etc.) on the removal performance, and explained the removal mechanism of the GO membrane on different antibiotics. The crosslinked EDA-GO membrane exhibited better stability and better removal performances on antibiotics than the non-crosslinked GO membrane. Na^+ in water could increase the flux of the EDA-GO membrane slightly but had no significant effect on the removal of antibiotics. Ca^{2+} could reduce the membrane flux and improve the removal of antibiotics. Mg^{2+} had little effect on the membrane flux and antibiotic removal. In the range of pH 6–9, with the increase of pH, the membrane flux of CAP and PNC solution increased, and the removal of CAP decreased, while the removal of PNC increased. The membrane flux and removal of ERY were not significantly affected by pH change. The removal efficiencies of low-concentration ($500\ \mu g \cdot L^{-1}$) antibiotic solutions were higher than those of high-concentration ($10\ mg \cdot L^{-1}$) antibiotic solutions. Furthermore, compared with the antibiotic solutions prepared with UP water, when filtering actual wastewater samples, J/J_{UP} of the membrane was lower and the removal of antibiotics was higher. In short, the EDA-GO membrane has a good application prospect in the removal of trace organic matters in water.

ACKNOWLEDGEMENTS

This work was supported by the National Natural Science Foundation of China (No. 21707100).

DATA AVAILABILITY STATEMENT

All relevant data are included in the paper or its Supplementary Information.

CONFLICT OF INTEREST

The authors declare there is no conflict.

REFERENCES

Bellona, C., Drewes, J. E., Xu, P. & Amy, G. 2004 Factors affecting the rejection of organic solutes during NF/RO treatment – a literature review. *Water Research* **38** (12), 2795–2809.

Choi, J. H., Jegal, J. & Kim, W. N. 2006 Fabrication and characterization of multi-walled carbon nanotubes/polymer blend membranes. *Journal of Membrane Science* **284** (1–2), 406–415.

Chu, K. H., Fathizadeh, M., Yu, M., Flora, J. R. V., Jang, A., Jang, M., Park, C. M., Yoo, S. S., Her, N. & Yoon, Y. 2017 Evaluation of removal mechanisms in a graphene oxide-coated ceramic ultrafiltration membrane for retention of natural organic matter, pharmaceuticals, and inorganic salts. *ACS Applied Materials & Interfaces* **9** (46), 40369–40377.

Dong, G. Y., Zhang, Y. T., Hou, J. W., Shen, J. N. & Chen, V. 2016 Graphene oxide nanosheets based novel facilitated transport membranes for efficient CO_2 capture. *Industrial & Engineering Chemistry Research* **55** (18), 5403–5414.

Ganesh, B. M., Isloor, A. M. & Ismail, A. F. 2013 Enhanced hydrophilicity and salt rejection study of graphene oxide-polysulfone mixed matrix membrane. *Desalination* **313**, 199–207.

Han, J. L., Xia, X., Tao, Y., Yun, H., Hou, Y. N., Zhao, C. W., Luo, Q., Cheng, H. Y. & Wang, A. J. 2016 Shielding membrane surface carboxyl groups by covalent-binding graphene oxide to improve anti-fouling property and the simultaneous promotion of flux. *Water Research* **102**, 619–628.

Hao, R. J., Zhang, Y., Du, T. T., Yang, L., Adeleye, A. S. & Li, Y. 2017 Effect of water chemistry on disinfection by-product formation in the complex surface water system. *Chemosphere* **172**, 384–391.

Hu, M. & Mi, B. X. 2013 Enabling graphene oxide nanosheets as water separation membranes. *Environmental Science & Technology* **47** (8), 3715–3723.

Huang, H. B., Mao, Y. Y., Ying, Y. L., Liu, Y., Sun, L. W. & Peng, X. S. 2013 Salt concentration, pH and pressure controlled separation of small molecules through lamellar graphene oxide membranes. *Chemical Communications* **49** (53), 5963–5965.

Hung, W. S., Tsou, C. H., De Guzman, M., An, Q. F., Liu, Y. L., Zhang, Y. M., Hu, C. C., Lee, K. R. & Lai, J. Y. 2014 Cross-linking with diamine monomers to prepare composite graphene oxide-framework membranes with varying d-spacing. *Chemistry of Materials* **26** (9), 2983–2990.

Kang, Y., Obaid, M., Jang, J., Ham, M. H. & Kim, I. S. 2018 Novel sulfonated graphene oxide incorporated polysulfone nanocomposite membranes for enhanced-performance in ultrafiltration process. *Chemosphere* **207**, 581–589.

Karunakaran, M., Villalobos, L. F., Kumar, M., Shevate, R., Akhtar, F. H. & Peinemann, K. V. 2017 Graphene oxide doped ionic liquid ultrathin composite membranes for efficient CO_2 capture. *Journal of Materials Chemistry A* **5** (2), 649–656.

Khanzada, N. K., Farid, M. U., Kharraz, J. A., Choi, J., Tang, C. Y., Nghiem, L. D., Jang, A. & An, A. K. 2020 Removal of organic micropollutants using advanced membrane-based water and wastewater treatment: a review. *Journal of Membrane Science* **598**, 117672.

Kong, F. X., Yang, H. W., Wang, X. M. & Xie, Y. F. F. 2014 Rejection of nine haloacetic acids and coupled reverse draw solute permeation in forward osmosis. *Desalination* **341**, 1–9.

Kong, F. X., Yang, H. W., Wu, Y. Q., Wang, X. M. & Xie, Y. F. F. 2015 Rejection of pharmaceuticals during forward osmosis and prediction by using the solution-diffusion model. *Journal of Membrane Science* **476**, 410–420.

Kong, F. X., Yang, H. W., Wang, X. M. & Xie, Y. F. F. 2016 Assessment of the hindered transport model in predicting the rejection of trace organic compounds by nanofiltration. *Journal of Membrane Science* **498**, 57–66.

Kong, F. X., Liu, Q., Dong, L. Q., Zhang, T., Wei, Y. B., Chen, J. F., Wang, Y. & Guo, C. M. 2020 Rejection of pharmaceuticals by graphene oxide membranes: role of crosslinker and rejection mechanism. *Journal of Membrane Science* **612**, 118338.

Lecaros, R. L. G., Mendoza, G. E. J., Hung, W. S., An, Q. F., Caparanga, A. R., Tsai, H. A., Hu, C. C., Lee, K. R. & Lai, J. Y. 2017 Tunable interlayer spacing of composite graphene oxide-framework membrane for acetic acid dehydration. *Carbon* **123**, 660–667.

Liang, B., Zhang, P., Wang, J. Q., Qu, J., Wang, L. F., Wang, X. X. & Guan, C. F. 2016 Membranes with selective laminar nanochannels of modified reduced graphene oxide for water purification. *Carbon* **103**, 94–100.

Liu, N., Jin, X. W., Feng, C. L., Wang, Z. J., Wu, F. C., Johnson, A. C., Xiao, H. X., Hollert, H. & Giesy, J. P. 2020 Ecological risk assessment of fifty pharmaceuticals and personal care products (PPCPs) in Chinese surface waters: a proposed multiple-level system. *Environment International* **136**, 105454.

Mahlangu, T. O., Schoutteten, K. V. K. M., D'haese, A., van den Bussche, J., Vanhaecke, L., Thwala, J. M., Mamba, B. B. & Verliefde, A. R. D. 2016 Role of permeate flux and specific membrane-foulant-solute affinity interactions (Delta G(slm)) in transport of trace organic solutes through fouled nanofiltration (NF) membranes. *Journal of Membrane Science* **518**, 203–215.

Meng, N., Zhao, W., Shamsaei, E., Wang, G., Zeng, X. K., Lin, X. C., Xu, T. W., Wang, H. T. & Zhang, X. W. 2018 A low-pressure GO nanofiltration membrane crosslinked via ethylenediamine. *Journal of Membrane Science* **548**, 363–371.

Mi, B. 2014 Graphene oxide membranes for ionic and molecular sieving. *Science* **343** (6172), 740–742.

Mo, Y. H., Zhao, X. & Shen, Y. X. 2016 Cation-dependent structural instability of graphene oxide membranes and its effect on membrane separation performance. *Desalination* **399**, 40–46.

Nam, Y. T., Choi, J., Kang, K. M., Kim, D. W. & Jung, H. T. 2016 Enhanced stability of laminated graphene oxide membranes for nanofiltration via interstitial amide bonding. *ACS Applied Materials & Interfaces* **8** (40), 27376–27382.

Nghiem, L. D., Schafer, A. I. & Elimelech, M. 2004 Removal of natural hormones by nanofiltration membranes: measurement, modeling, and mechanisms. *Environmental Science & Technology* **38** (6), 1888–1896.

Qi, H. B., Zhao, X., Li, H., Che, Y. & Wang, C. 2020 Impact of monovalent cations on the separation performance of graphene oxide membrane for different organic matters. *Water Science & Technology* **82** (8), 1560–1569.

Song, N., Gao, X. L., Ma, Z., Wang, X. J., Wei, Y. & Gao, C. J. 2018 A review of graphene-based separation membrane: materials, characteristics, preparation and applications. *Desalination* **437**, 59–72.

Steinle-Darling, E., Litwiller, E. & Reinhard, M. 2010 Effects of sorption on the rejection of trace organic contaminants during nanofiltration. *Environmental Science & Technology* **44** (7), 2592–2598.

Verliefde, A. R. D., Cornelissen, E. R., Heijman, S. G. J., Hoek, E. M. V., Amy, G. L., van der Bruggen, B. & van Dijk, J. C. 2009 Influence of solute-membrane affinity on rejection of uncharged organic solutes by nanofiltration membranes. *Environmental Science & Technology* **43** (7), 2400–2406.

Wallmann, L., Krampe, J., Lahnsteiner, J., Radu, E., van Rensburg, P., Slipko, K., Wogerbauer, M. & Kreuzinger, N. 2021 Fate and persistence of antibiotic-resistant bacteria and genes through a multi-barrier treatment facility for direct potable reuse. *Water Reuse* **11** (3), 373–390.

Wang, X. L., Li, Y. L., Yu, H. T., Yang, F. L., Tang, C. Y., Quan, X. & Dong, Y. C. 2020 High-flux robust ceramic membranes functionally decorated with nano-catalyst for emerging micro-pollutant removal from water. *Journal of Membrane Science* **611**, 118281.

Xu, Y., Li, Z., Su, K., Fan, T. T. & Cao, L. 2018 Mussel-inspired modification of PPS membrane to separate and remove the dyes from the wastewater. *Chemical Engineering Journal* **341**, 371–382.

Yeh, C. N., Raidongia, K., Shao, J., Yang, Q. H. & Huang, J. X. 2015 On the origin of the stability of graphene oxide membranes in water. *Nature Chemistry* **7** (2), 166–170.

Ying, Y. P., Liu, D. H., Ma, J., Tong, M. M., Zhang, W. X., Huang, H. L., Yang, Q. Y. & Zhong, C. L. 2016 A GO-assisted method for the preparation of ultrathin covalent organic framework membranes for gas separation. *Journal of Materials Chemistry A* **4** (35), 13444–13449.

Yu, Z. Y., Song, Z. H., Wen, X. H. & Huang, X. 2015 Using polyaluminum chloride and polyacrylamide to control membrane fouling in a cross-flow anaerobic membrane bioreactor. *Journal of Membrane Science* **479**, 20–27.

Yuan, B. Q., Wang, M. X., Wang, B., Yang, F. L., Quan, X., Tang, C. Y. & Dong, Y. C. 2020 Cross-linked graphene oxide framework membranes with robust nano-channels for enhanced sieving ability. *Environmental Science & Technology* **54** (23), 15442–15453.

Zhang, P., Gong, J. L., Zeng, G. M., Deng, C. H., Yang, H. C., Liu, H. Y. & Huan, S. Y. 2017a Cross-linking to prepare composite graphene oxide-framework membranes with high-flux for dyes and heavy metal ions removal. *Chemical Engineering Journal* **322**, 657–666.

Zhang, H. J., Li, B., Pan, J. F., Qi, Y. W., Shen, J. N., Gao, C. J. & Van der Bruggen, B. 2017b Carboxyl-functionalized graphene oxide polyamide nanofiltration membrane for desalination of dye solutions containing monovalent salt. *Journal of Membrane Science* **539**, 128–137.

Zhao, J., Zhu, Y. W., He, G. W., Xing, R. S., Pan, F. S., Jiang, Z. Y., Zhang, P., Cao, X. Z. & Wang, B. Y. 2016 Incorporating zwitterionic graphene oxides into sodium alginate membrane for efficient water/alcohol separation. *ACS Applied Materials & Interfaces* **8** (3), 2097–2103.

Zhao, J. J., Yang, Y., Li, C., Liang, J. & Hou, L. A. 2019 Impacts of mono/divalent cations on the lamellar structure of cross-linked GO layers and membrane filtration performance for different DOM fractions. *Chemosphere* **237**, 124544.

First received 28 November 2022; accepted in revised form 19 April 2023. Available online 6 May 2023

doi: 10.2166/wrd.2023.081

Fenton process enhanced by metal sulfide for treating the actual evaporated mother liquid of gas field wastewater

Bing Yao, Ying Chen, Mengzhe Wang and Min Liu*

College of Architecture and Environment, Sichuan University, Chengdu 610065, China
*Corresponding author. E-mail: liuminscu@163.com

ABSTRACT

Evaporated mother liquor of gas field wastewater (EML-GFW) is a form of wastewater generated by the triple-effect evaporation treatment of gas field wastewater containing complex pollutants. In this study, four metal sulfides, CuS, ZnS, MoS_2, and WS_2, were used to strengthen the Fenton process in EML-GFW treatment. The optimum Fenton/ZnS process for the highest removal of TOC from EML-GFW was achieved at the initial pH of 3.0 and in a mixture of $FeSO_4·7H_2O$:ZnS:H_2O_2 in the ratio of 30 g/L:10 g/L:1.2 mol/L, with a TOC removal efficiency of 74.5%. The organic components analysis of EML-GFW over four distinct periods demonstrated that the presence of N,N-dimethylethanolamine (DMEA) persisted and accounted for the greatest proportion of pollutants, identifying it as the characteristic pollutant. The TOC removal mechanism by Fenton/ZnS was revealed via analysis of organic materials obtained from the Fenton/ZnS process, tert-butanol quenching experiment, and illumination experiment. ZnS-generated hole–electron pairs under illumination, which promoted the reduction of Fe^{3+} to Fe^{2+}, followed by an acceleration of •OH generation, thus improving TOC removal efficiency. The Fenton/ZnS process improved the treatment of EML-GFW in the laboratory, providing strong data support and theoretical guidance for expanding this technology at the gas field project site.

Key words: evaporated mother liquor, Fenton/ZnS, gas field wastewater, N, N-dimethylethanolamine, total organic carbon (TOC)

HIGHLIGHTS

- The optimization of the Fenton/ZnS process was based on the actual wastewater.
- The TOC removal efficiency of EML-GFW reached 74.5% by the optimum Fenton/ZnS process.
- ZnS accelerated the production of •OH, increasing the removal efficiency of TOC.
- This work was meaningful for the wastewater treatment at the gas field project site.

GRAPHICAL ABSTRACT

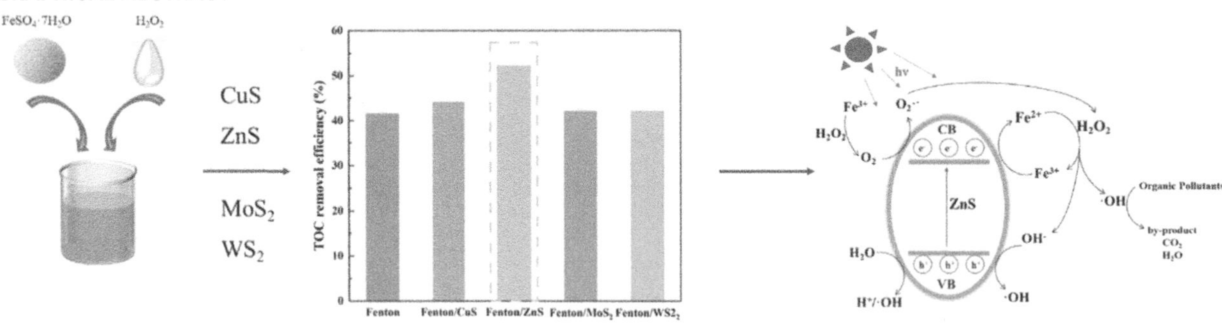

Fenton   Fenton processes enhanced by mental sulfides   TOC removal mechanism by Fenton/ZnS

1. INTRODUCTION

Natural gas is considered low-carbon clean energy that ranks between traditional fossil fuels, such as oil and coal, and renewable energy. Natural gas emits less carbon than coal and oil and is cheaper than renewable energy, making it a suitable platform to foster China's energy transition toward achieving effective air pollution control and carbon neutrality (Ibrahim *et al.* 2022). China's natural gas production has increased recently, reaching 1,925.0 billion m^3 at the end of 2020 (Resources 2021). However, during natural gas extraction, a huge volume of gas field wastewater is generated, including drilling wastewater, fracturing flow-back fluid, and high-salt wastewater (Feng *et al.* 2020; Jin *et al.* 2022). In addition, the triple-effect evaporation process has been identified as an effective technique for treating gas field wastewater due to the operability of evaporation equipment at the site of the gas field project. However, the evaporated mother liquor of gas field wastewater (EML-GFW), the concentrated liquid remaining after distillation, contains a range of organic compounds and inorganic salt ions at significantly higher concentrations than raw water (Wang *et al.* 2019; Feng *et al.* 2022).

There are only a handful of studies conducted on the actual EML-GFW due to the complexity of actual wastewater. EML-GFW removal using traditional adsorption methods, air flotation separation, and coagulation could only transform the phases of pollutants, thus necessitating additional treatment. In addition, a high-salt concentration leads to membrane fouling, thermal equipment scaling, and microbial death, thus hindering the application of these methods in EML-GFW treatment (Moussa *et al.* 2006; Osaka *et al.* 2008; Sheng *et al.* 2022).

The Fenton method is an established advanced oxidation technique (AOP) used to degrade refractory organic pollutants due to its strong oxidizing ability, fast reaction speed, and high treatment efficiency (Adil *et al.* 2020; Wang *et al.* 2021; Ziembowicz & Kida 2022). However, the decomposition efficiency of H_2O_2 in conventional Fenton is limited because of the weak cyclic reaction of Fe^{2+}/Fe^{3+}, which requires a catalyst or activator to accelerate Fe^{2+} regeneration (Liu *et al.* 2018; Ren *et al.* 2022). Metal sulfides are abundant in nature and have good electrical conductivity and strong stability, making them an ideal catalyst candidate to accelerate the speed and efficiency of pollutant removal by AOP (Fu & Lee 2019; Zhu *et al.* 2020; Li *et al.* 2021; Kumar *et al.* 2022). A diversity of metal sulfides have been reported to act as excellent catalysts to enhance ·OH production, including CuS, MoS_2, WS_2, ZnS, Cr_2S_3, CoS_2, PbS, and so on. Xing *et al.* used the Fenton methods enhanced by various metal sulfides (MoS_2, WS_2, ZnS, Cr_2S_3, CoS_2, and PbS) to treat 20 mg/L rhodamine B (RhB). The results revealed that not only had the removal efficiency of RhB reached more than 90%, but also the reaction rate constant of RhB removal by the Fenton/MoS_2 method (3.7×10^{-2}/s) was 18.5 times higher than that of the conventional Fenton method (0.2×10^{-2}/s) (Xing *et al.* 2018). Dong *et al.* reported that in the WS_2 co-catalytic photoassisted Fe(II)/H_2O_2 Fenton system, accompanied by improved H_2O_2 decomposition efficiency, it was able to oxidize phenol (10 mg/L) and reduce Cr(VI) (40 mg/L) simultaneously, with removal efficiencies of 80.9 and 90.9%, respectively (Dong *et al.* 2018).

Four metal sulfides (CuS, ZnS, MoS_2, and WS_2) were used to enhance the Fenton process and to determine the process with the highest efficiency for actual EML-GFW. Based on the fundamental factors, namely the reaction time, initial pH of the solution, and the dosage of H_2O_2 and metal sulfides, optimization of the metal sulfide-enhanced process was conducted. Finally, simulated wastewater of characteristic pollutants, *N,N*-dimethylethanolamine (DMEA), was prepared to explore the TOC removal mechanism by Fenton/ZnS without the interference of complex organic matters in actual EML-GFW. This research offers theoretical insight into the implementation of the Fenton process, particularly for treating organic wastewater with complex constituents.

2. MATERIALS AND METHODS

2.1. Materials

EML-GFW was collected from a gas production plant in the Northeast Sichuan Province, China, where the gas field wastewater was first subjected to a pre-treatment process, followed by a treatment using the triple-effect evaporation technology to obtain EML-GFW. In addition to its poor biodegradability (B/C = 0.20), EML-GFW was found to contain salts in a high concentration. In Figure 1, the organic components analysis of EML-GFW differs substantially between the first batch (stored for 2 years), second batch, and third batch of EML-GFW, except for one common finding: the presence of DMEA as one of the major constituents in all batches. While the organic matters in EML-GFW typically react and transform with one another, the primary pollutant, DMEA, proved difficult to convert. Therefore, DMEA was identified as a characteristic organic pollutant of EML-GFW. Considering the complexity of actual EML-GFW, DMEA-simulated wastewater (DSW) was created to explore the removal mechanism of organic pollutants. In addition, all actual EML-GFW was obtained from the freshest third batch for all experiments conducted, and water quality is presented in Table 1.

Sulfuric acid (H_2SO_4), sodium hydroxide (NaOH), hydrogen peroxide (H_2O_2, 30%, w/w), ferrous sulfate heptahydrate ($FeSO_4·7H_2O$), tert-butanol (($CH_3)_3COH$), copper sulfide (CuS), molybdenum disulfide (MoS_2), zinc sulfide (ZnS), tungsten

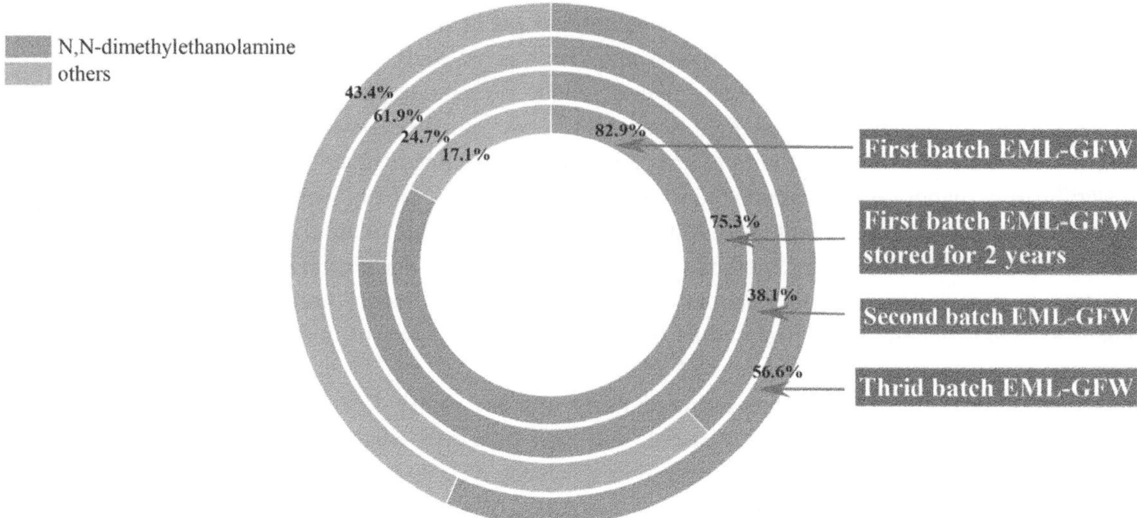

Figure 1 | The organic matter compositions of the first batch, first batch (stored for 2 years), second batch, and third batch EML-GFW wastewater by GC-MS spectra.

Table 1 | Indicators for water quality of EML-GFW for all the experiments conducted

Characteristics	Unit	Value
pH	–	13.5
Chemical oxygen demand (COD)	mg/L	1.12×10^4
Biochemical oxygen demand (BOD_5)	mg/L	2.21×10^3
Ammonia nitrogen (N)	mg/L	1.12
Suspended solids (SS)	mg/L	2.14×10^2
Petroleum	mg/L	5.80×10^{-1}
Volatile phenols (Phenol)	mg/L	9.00×10^{-2}
Total alkalinity ($CaCO_3$)	mg/L	4.92×10^4
Chloride	mg/L	8.50×10^4
Sulfate	mg/L	1.44×10^4
Conductivity	μS/cm	1.60×10^3
TOC	mg/L	1.55×10^3

disulfide (WS_2), dimethylamine (($CH_3)_2NH$), N,N-dimethylethanolamine (($CH_3)_2NCH_2CH_2OH$) used were all of the analytically pure (AR) grade. Distilled water was used in the experiments.

2.2. Experimental procedures

2.2.1. Batch experiments on the TOC removal effect of EML-GFW

The pH of 100 mL of EML-GFW was adjusted to 4.0 with sulfuric acid and subsequently added with 22 g/L $FeSO_4 \cdot 7H_2O$ and 20 g/L metal sulfides (CuS, ZnS, MoS_2, or WS_2) with vigorous stirring. After a reaction with 0.8 mol/L H_2O_2 for 3 h, the Fenton reaction was terminated. The supernatant of the precipitated solution was passed through a filter with a pore of 0.45 μm in diameter to measure TOC, followed by the screening of metal sulfides with the best treatment efficiency. Then, the effects of reaction time (1, 2, 3, 4, 5, and 6 h), initial pH of the solution (2.0, 3.0, 4.0, 5.0, 6.0, and 7.0), the dosage of H_2O_2 (0.2, 0.4, 0.6, 0.8, 1.0, 1.2, and 1.4 mol/L), $FeSO_4 \cdot 7H_2O$ (10, 15, 20, 25, 30, 35, and 40 g/L) and metal sulfide (0, 2, 4, 6, 8, 10, and 12 g/L) on EML-GFW were investigated. All experiments were performed three times to ensure the accuracy and reliability of the results.

2.2.2. Identification of active substances in DSW treatment

Based on DMEA concentration in the actual wastewater (EML-GFW), the simulated wastewater (DSW) was prepared with 386 mg/L DMEA. Fenton/ZnS process dosages were adjusted in proportions equal to the TOC ratios of EML-GFW and DSW.

The pH of DSW was adjusted to 3.0, and 1 mol/L tert-butanol was added. After fully dissolving 4.6 g/L $FeSO_4 \cdot 7H_2O$ and 1.5 g/L metal sulfide in DSW, 0.185 mol/L H_2O_2 was slowly added. Following the termination of the Fenton reaction, the supernatant was passed through a filter membrane with a pore size of 0.45 µm and subjected to DMEA concentration determination.

2.2.3. Effect of light on DSW treatment

The foil wrap was used to block the light and the next steps were performed according to Section 2.2.2. All experiments were conducted three times to ensure the accuracy and reliability of the results.

2.3. Analytical methods

TOC measurement (vario TOC select, Germany) was conducted according to the combustion oxidation-non-dispersive infrared absorption method. pH was measured by a microelectrode system (unisense mmm 7221, Danish).

EML-GFW samples were subjected to solid-phase micro-extraction and analyzed by gas chromatography-mass spectrometry (GC-MS) to qualitatively determine the compositions of organic compounds. GC-MS analysis was carried out using the following protocols: DB-624UI (60 m × 0.25 mm × 1.40 µm) column; temperature procedures: 40 °C for 5 min, followed by a gradual increment to 150 °C at a rate of 5 °C/min, the final increment to 280 °C at a rate of 10 °C/min, and a holding time of 7 min; high-purity helium as a carrier gas; a diversion ratio of 3.0. The full scan mode was adopted at the scan range of 33–500 amu. Finally, the retention time of the compound was compared with the NIST database to determine the quality of the results. These tests were completed at the Analytical Testing Center of Sichuan University.

DMEA concentration was determined by gas chromatography using the following parameters: CAM (30 m × 0.32 mm × 0.25 µm) column; temperature procedure: 60 °C for 5 min, followed by the increment to 200 °C at a rate of 20 °C/min, and a holding time of 4 min. The injection volume was 1 µL and nitrogen was used as the carrier gas at a flow rate of 25 cm/s.

3. RESULTS AND DISCUSSION

3.1. TOC removal effect of EML-GFW by the enhanced Fenton process

To determine the capacity of metal sulfides in enhancing the Fenton process, five experiments were conducted by adding CuS, ZnS, MoS_2, or WS_2 into an EML-GFW solution as described in Figure 2. The TOC removal efficiencies of the Fenton process enhanced with the addition of CuS, ZnS, MoS_2, or WS_2 were 41.6, 44.1, 52.3, and 42.1%, respectively. The results depict that the highest enhancement of efficiency of the Fenton process was achieved by adding ZnS, possibly attributed to the photocatalytic activity of ZnS. Unlike the classical Fenton process, the presence of a semiconductor material, ZnS, was capable of acting as a catalyst by absorbing the visible light with energy greater than or equal to its band gap, causing the excitation of valence band electrons in the conduction band. The electrons and holes induced by light could further generate ·OH as displayed in Equations (9)–(13, followed by redox contaminants, consistent with the results of Xing *et al.* (Wang *et al.* 2008; Lee & Wu 2017; Xing *et al.* 2018; Manny Porto Barros *et al.* 2023). In addition, the Fe^{3+} generated during the Fenton process could accept the electrons generated from the visible light irradiation of ZnS. Enhancing the Fe^{3+}/Fe^{2+} cycle can effectively ensure H_2O_2 decomposition and thus, the continuous generation of free radicals, avoiding the blockage after complete depletion of Fe^{2+}. Simultaneously, it effectively reduced the compounding of photogenerated electrons with holes (Guo *et al.* 2022; Zhang & Chu 2022). The photocatalytic activities of CuS, MoS_2, and WS_2 were weaker than those of ZnS, resulting in less effective treatment.

3.2. Optimization of the Fenton/ZnS process

The optimization of the Fenton/ZnS process was conducted based on several factors: reaction time, initial solution pH, as well as the dosages of H_2O_2, $FeSO_4 \cdot 7H_2O$, and ZnS.

3.2.1. Reaction time

Figure 3(a) displays the effect of reaction time within 6 h on the TOC removal by ZnS-enhanced Fenton. In the first 3 h, TOC removal efficiency increased slightly and finally reached 52.8%. After 6 h of reaction, TOC removal efficiency slightly increased

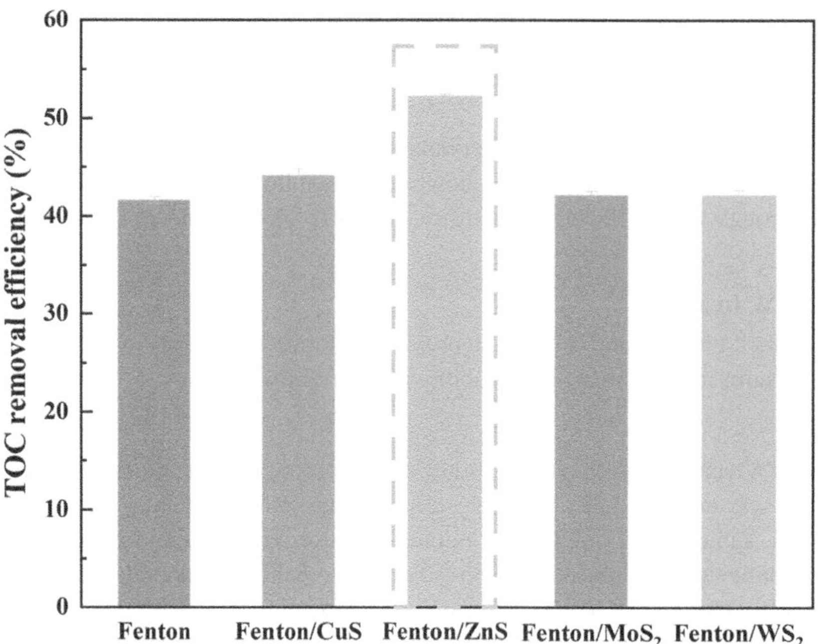

Figure 2 | Effect of different metal sulfides on TOC removal efficiency of EML-GFW. (The initial pH was 4.0; the concentration of FeSO$_4$·7H$_2$O, metal sulfides, and H$_2$O$_2$ was 22, 20, and 0.8 mol/L, respectively; the reaction time was 3 h).

to 53.5%, without significant improvement. At the initial stage of the reaction, Fe^{2+} reacted with H$_2$O$_2$ to generate ·OH as described by Equation (1), which could effectively mineralize pollutants, and Fe^{3+} produced during this period was reduced to Fe^{2+} according to Equations (2) and (3) due to ZnS photocatalysis, resulting in a decreased TOC concentration (Chen *et al.* 2019; Guo *et al.* 2021). However, in addition to the low rate constant of the reaction as illustrated in Equation (2), the active sites of ZnS in EML-GFW might be masked by impurities, causing Fe^{3+} to accumulate in significant amounts while the catalyst, Fe^{2+}, continued to decrease as the reaction progressed, lowering the rate of mineralization. Extending the reaction time could not improve TOC removal; consequently, 3 h is the optimum reaction time for the enhanced Fenton process.

$$Fe^{2+} + H_2O_2 \rightarrow Fe^{3+} + OH^- + \bullet OH \tag{1}$$

$$Fe^{3+} + H_2O_2 \rightarrow Fe^{2+} + HO_2\bullet + H^+ \tag{2}$$

$$Fe^{3+} + HO_2\bullet \rightarrow Fe^{2+} + O_2 + H^+ \tag{3}$$

3.2.2. Initial pH of the solution

Generally, the oxidation in the Fenton process was found to be highest under acidic conditions; accordingly, the Fenton/ZnS process for treating EML-GFW was conducted using initial pH values of 2.0, 3.0, 4.0, 5.0, 6.0, and 7.0 for 3 h (Figure 3(b)). When pH increased from 2.0 to 3.0, TOC efficiency increased from 48.4 to 55.2%; however, a further increase to 7.0 significantly decreased the removal efficiency to 37.3%. Due to the better stability of solid ZnS at pH 3.0–11, the initial pH affected pollutant removal mainly by influencing the Fe^{2+}-catalyzed Fenton reaction, so the removal efficiency was higher under acidic than neutral conditions (Wang *et al.* 2011; Manny Porto Barros *et al.* 2023). H$^+$ in the solution exhibited a strong scavenging effect on ·OH according to Equation (4) at pH 2.0 and a low TOC removal effect at pH 3.0. As pH continued to increase, Fe^{3+} in the solution was hydrolyzed and precipitated, which affected the catalytic ability of the catalyst, resulting in a significant decline in the treatment efficiency (Huang & Huang 2008; Navalon *et al.* 2011). The optimal initial reaction pH for the Fenton/ZnS process was 3.0.

$$\bullet OH + H^+ + e^- \rightarrow H_2O \tag{4}$$

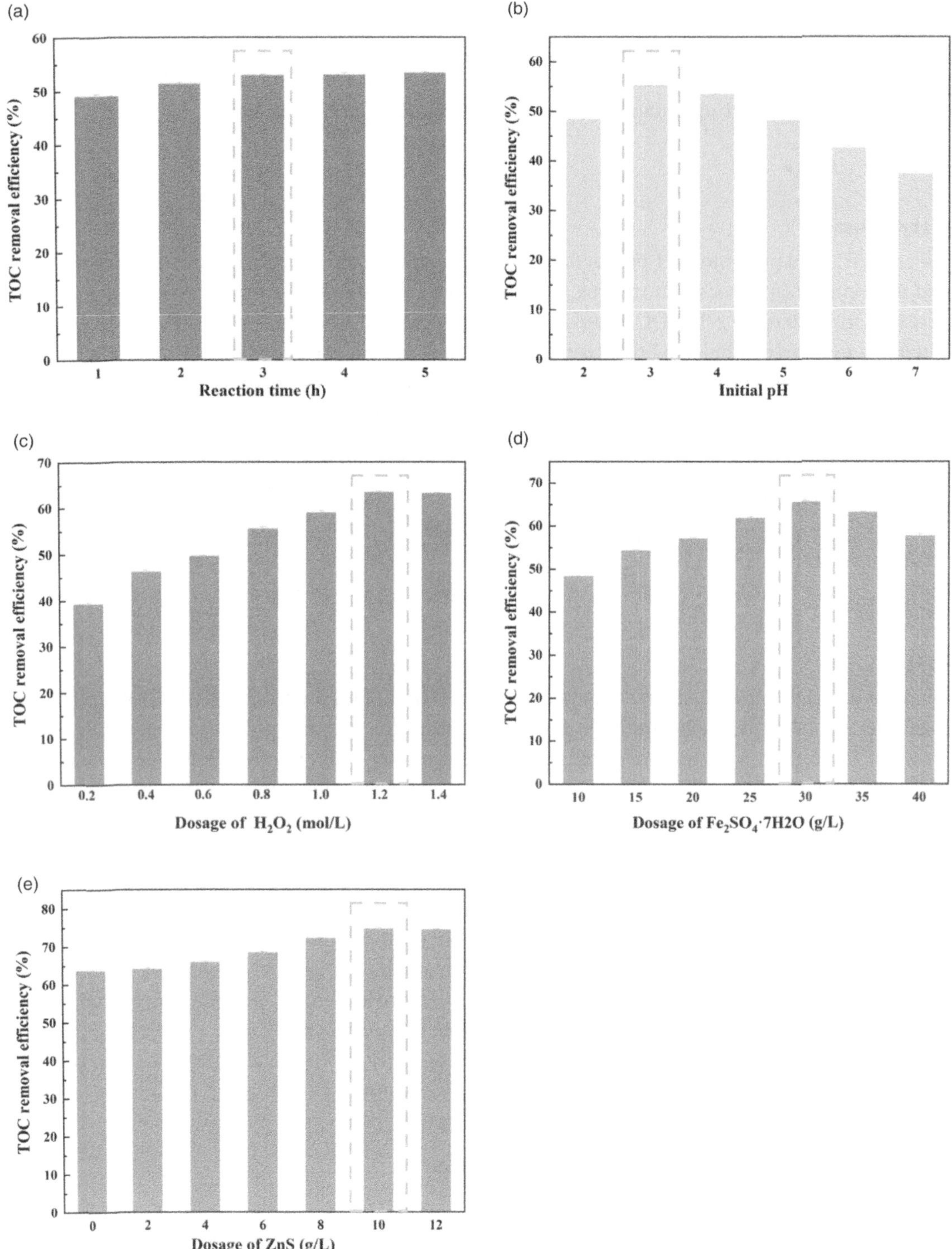

Figure 3 | Effect of reaction time (a); initial pH of the solution (b); dosage of H_2O_2 (c); $FeSO_4 \cdot 7H_2O$ (d); and ZnS (e) on TOC removal of EML-GFW.

3.2.3. H_2O_2 dosage

H_2O_2 is the main contributor of hydroxyl free radicals and the effect of its dosage on TOC removal via the Fenton/ZnS process at pH 3.0 after 3 h of reaction is demonstrated in Figure 3(c). When H_2O_2 dosage was increased from 0.2 to 1.2 mol/L, the removal efficiency of TOC increased from 39.2 to 63.3% as more •OH was produced to mineralize the organic matter.

However, when H_2O_2 dosage was further increased to 1.4 mol/L, the quenching effect on •OH would change to a state as described in Equation (5), which results in ineffective decomposition and is not conducive to the mineralization of organic compounds by •OH (Ramirez *et al.* 2007; Xue *et al.* 2009a, 2009b; Jiang *et al.* 2010). Adding excessive H_2O_2 would not only render the process expensive but also ineffective in improving the removal efficiency of TOC. Consequently, it was decided that 1.2 mol/L H_2O_2 is sufficient for optimum TOC removal via the Fenton/ZnS process.

$$H_2O_2 + \bullet OH \rightarrow H_2O + HO_2 \bullet \tag{5}$$

3.2.4. FeSO$_4$·7H$_2$O dosage

Using Fe^{2+} as a catalyst could promote •OH production by H_2O_2, and thus the effect of different dosages of FeSO$_4$·7H$_2$O on the efficiency of the Fenton/ZnS process in treating EML-GFW was investigated as described in Figure 3(d). When FeSO$_4$·7H$_2$O was increased from 10 to 30 g/L, TOC removal efficiency increased from 48.3 to 65.5%; however, a further increase to 40 g/L resulted in a decreased value to 57.6%. Using FeSO$_4$·7H$_2$O at concentrations higher than 30 g/L reduced the TOC removal efficiency. In addition to promoting •OH quenching by Fe^{2+} as shown in Equation (6), the excessive ferric produces several mutually quenched •OH, as displayed in Equation (7), resulting in the loss of available •OH reducing organic mineralization efficiency (Liu *et al.* 2015). Furthermore, excess FeSO$_4$·7H$_2$O increases the amount of dye added for subsequent pH recovery, thus increasing the cost of wastewater treatment. Therefore, the optimal dosage of FeSO$_4$·7H$_2$O for the Fenton/ZnS process was determined to be 30 g/L.

$$Fe^{2+} + \bullet OH \rightarrow Fe^{3+} + OH^- \tag{6}$$

$$\bullet OH + \bullet OH \rightarrow H_2O_2 \tag{7}$$

3.2.5. ZnS dosage

ZnS addition was beneficial to improve the treatment effect of the Fenton/ZnS process on EML-GFW and the influence of its dosage on wastewater treatment is demonstrated in Figure 3(e). With the increase of ZnS dosage, TOC removal efficiency exhibited a trend of increasing, followed by stabilizing. The maximum TOC removal efficiency was 74.5% when ZnS dosage was increased from 0 to 12 g/L. The photocatalytic activity of ZnS helped improve the oxidation activity of Fenton because ZnS contributed to Fe^{3+} reduction to Fe^{2+} when illuminated, and the regenerated Fe^{2+} continued to react with H_2O_2 to generate more •OH, thus promoting the continuous progress of the Fenton reaction and the mineralization of organic pollutants (Xing *et al.* 2018). However, when ZnS was added in excess, TOC removal effect was not improved due to the limited oxidant H_2O_2 content in the solution. Meanwhile, the excess ZnS led to turbidity, which hindered photogenerated radiation and subsequently decreased catalyst efficiency. In addition, the formation of particle agglomerates impaired the excitation process and the formation of electron/hole pairs due to ZnS overdosing (Suave *et al.* 2018; Garg *et al.* 2019; Manny Porto Barros *et al.* 2023). Compared with the single Fenton method, the Fenton/ZnS process with 10 g/L ZnS addition could improve the TOC removal efficiency of EML-GFW by 32.9%.

Under optimized conditions, i.e., initial pH of 3.0, 30 g/L of FeSO$_4$ 7H$_2$O, 10 g/L ZnS, and 1.2 mol/L H_2O_2, despite the presence of high salt and multiple organics in EML-GFW, TOC removal efficiency by the Fenton/ZnS process could reach up to 74.5% after 3 h of reaction.

3.3. Mechanism of DSW treatment by the Fenton/ZnS process

3.3.1. Analysis of organics after the Fenton/ZnS process

DMEA was undetected in EML-GFW and DSW subjected to the optimized Fenton/ZnS process and TOC removal efficiencies were recorded as 74.5 and 46.1%, respectively, indicating the easy removal yet difficult mineralization of DMEA. The main organic compounds identified in DSW subjected to the Fenton/ZnS process include dimethylamine, with a relative percentage of 90.8%, suggesting that part of DMEA might have completely mineralized to produce CO_2 and H_2O, and the rest was mainly oxidized to form dimethylamine, which remained in the solution.

3.3.2. Identification of active substances in DSW treatment by the Fenton/ZnS process

The potential rapid reaction between tert-butanol (TBA) and •OH at a rate of 6.0×10^8 L/(mol·s) is described by Equation (8), which is a quencher commonly used to demonstrate the presence of •OH (Witte *et al.* 2009; Lai *et al.* 2021). Based on Figure 4,

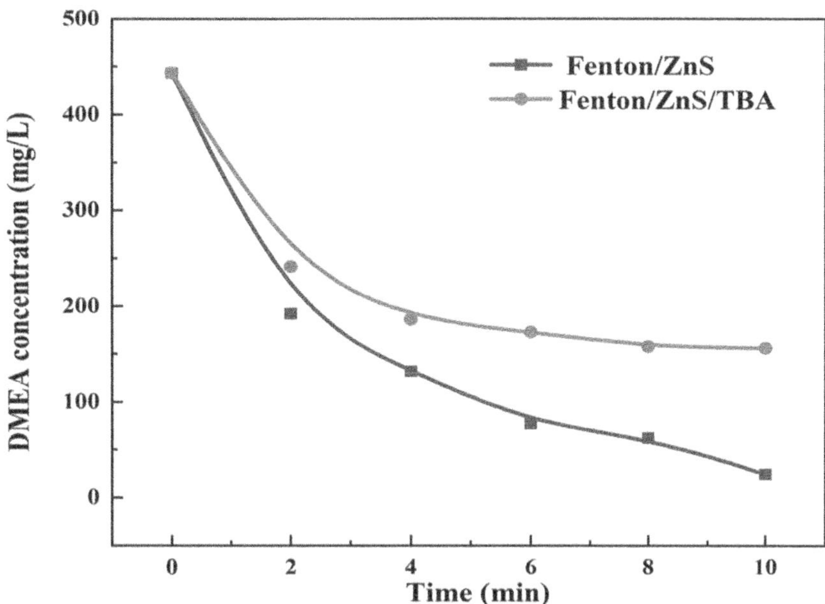

Figure 4 | Effect of TBA on DSW treatment via the optimized Fenton/ZnS process. (The initial pH was 3.0; the dosages of FeSO$_4$·7H$_2$O, H$_2$O$_2$, ZnS, and tert-butanol were 4.6 g/L, 0.185 mol/L, 1.5 g/L, and 1 mol/L, respectively; the reaction time was 10 min).

DMEA concentration decreased from 94.5 to 64.8% with TBA addition at 10 min, indicating a significant inhibition of DMEA removal by TBA in the Fenton/ZnS process, with •OH playing an important role in DSW mineralization.

$$(CH_3)_3COH + •OH \rightarrow H_2O + •CH_2(CH_3)_2COH \tag{8}$$

3.3.3. Effect of light on DSW treatment by the Fenton/ZnS process

ZnS is a semiconductor material often used in the photocatalytic treatment of wastewater (Zhang *et al.* 2011; Hao *et al.* 2018). Under shading and illumination, DSW treatment by the Fenton/ZnS is shown in Figure 5. TOC removal efficiency via the Fenton/ZnS process without shading was 46.1%, while under shading conditions, the efficiency was 27.3%, indicating the role of light in creating a conducive environment to DMEA mineralization via the Fenton/ZnS process. The possible mechanism of EML-GFW treatment via the Fenton/ZnS process is depicted in Figure 6. ZnS-generated electrons and holes under illumination as in Equation (9), both favoring the induction of free radicals. The photogenerated electrons reduced Fe^{3+} in the solution to Fe^{2+}, thereby promoting •OH generation in the Fenton reaction. Simultaneously, photogenerated electrons reacted with O$_2$ to produce •O$_2^-$ as expressed by Equation (10), generating H$_2$O$_2$ as expressed by Equation (11), which further supported the continuous progress of the Fenton reaction. In addition, photogenic holes might react with H$_2$O or OH$^-$ to produce •OH as described in Equations (12) and (13) (Rajeshwar *et al.* 2008; Zhu *et al.* 2019). •OH generated in these processes could oxidize DMEA to form by-products, dimethylamine, or even mineralizing to CO$_2$ and H$_2$O.

$$ZnS + h\nu \rightarrow ZnS(e^-) + ZnS(h^+) \tag{9}$$

$$ZnS(e^-) + O_2 \rightarrow •O_2^- \tag{10}$$

$$•O_2^- + ZnS(e^-) + 2H^+ \rightarrow H_2O_2 \tag{11}$$

$$ZnS(h^+) + H_2O \rightarrow H^+ + •OH \tag{12}$$

$$ZnS(h^+) + OH^- \rightarrow •OH \tag{13}$$

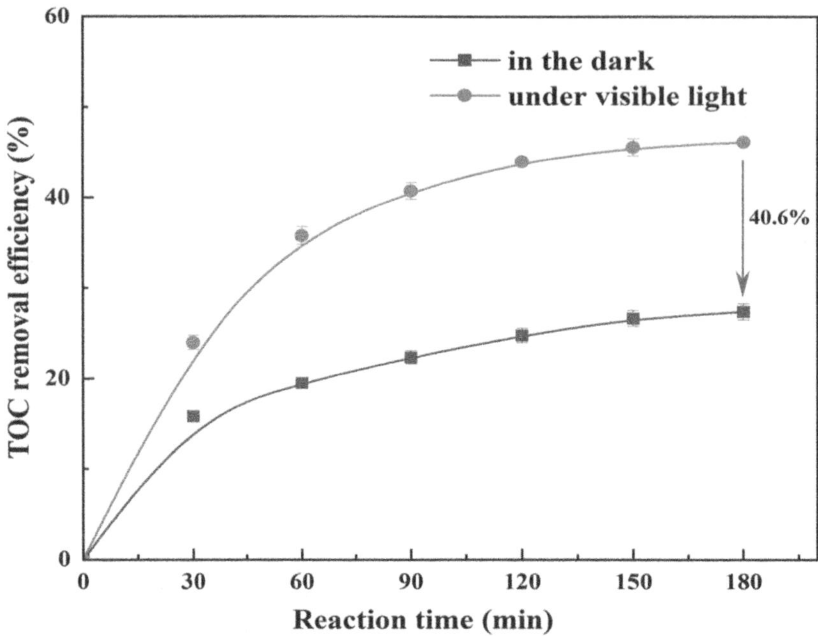

Figure 5 | Effect of light condition on DSW treatment via the optimized Fenton/ZnS process. (The initial pH was 3.0; the dosages of $FeSO_4·7H_2O$, H_2O_2, and ZnS were 4.6 g/L, 0.185 mol/L, and 1.5 g/L, respectively; the reaction time was 3 h).

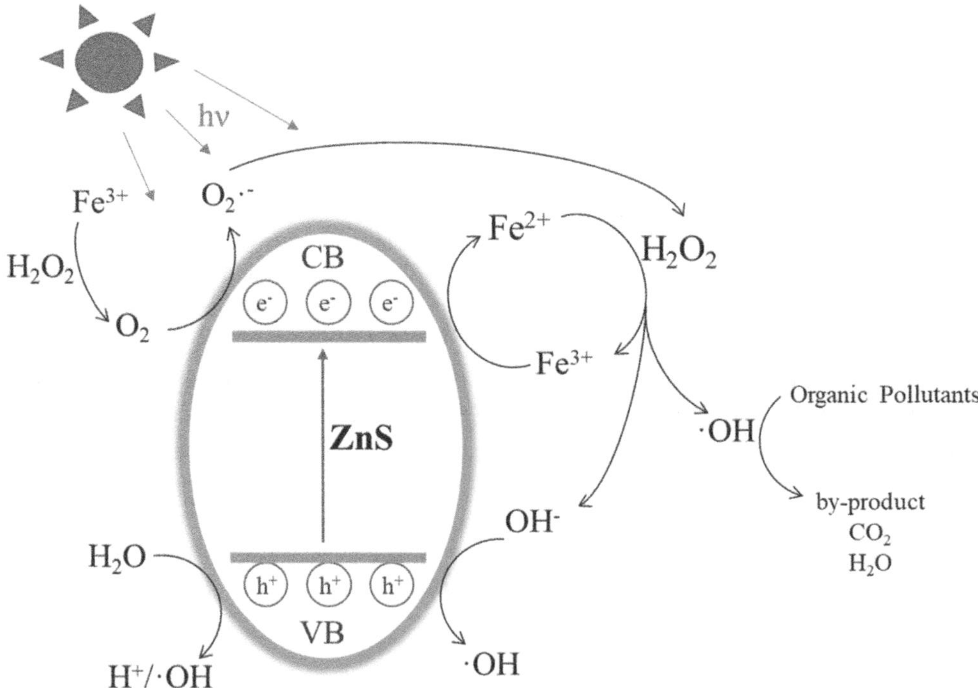

Figure 6 | Mechanism of EML-GFW treatment via the optimized Fenton/ZnS process. (The initial pH was 3.0; the dosages of $FeSO_4·7H_2O$, H_2O_2, and ZnS were 4.6 g/L, 0.185 mol/L, and 1.5 g/L, respectively; the reaction time was 3 h).

4. CONCLUSION

The effects of the Fenton process enhanced by metal sulfides (CuS, ZnS, MoS_2, and WS_2) on actual EML-GFW were investigated, revealing that the optimum condition was achieved by applying the Fenton/ZnS process. The TOC removal efficiency of EML-GFW reached 74.5% when the process was carried out with an initial pH of 3.0, and the doses of $FeSO_4 \cdot 7H_2O$, ZnS, and H_2O_2 used were 30, 10 g/L, and 1.2 mol/L, respectively. Furthermore, GC-MS analysis identified DMEA as one of the main pollutants of EML-GFW at different time points. Finally, the mechanism of DSW treatment via the Fenton/ZnS process was proposed. The photocatalytic activity of ZnS under illumination could accelerate the Fenton reaction by promoting the production of the active component, •OH, increasing the removal efficiency of TOC. In summary, the Fenton/ZnS process may allow for a better treatment efficiency of the actual wastewater EML-GFW composed of complex organic contaminants in the laboratory, offering valuable insights for application in wastewater treatment at the gas field project site.

AUTHOR CONTRIBUTIONS

B.Y. conceptualized the system, investigated the study, did a formal analysis, and wrote the original draft. Y.C. and M.L. supervised the study, wrote the review, edited the file, managed resources, and acquired funds. M.W. investigated the study, wrote the review and edited the file.

FUNDING

This study was supported by the National Major Science and Technology Project of the 13th Five-Year Plan 'High-efficiency development of ultra-deep bio-herm gas reservoirs with bottom water' of China (2016ZX05017-005).

DATA AVAILABILITY STATEMENT

All relevant data are included in the paper or its Supplementary Information.

CONFLICT OF INTEREST

The authors declare there is no conflict.

REFERENCES

Adil, S., Maryam, B., Kim, E.-J. & Dulova, N. 2020 Individual and simultaneous degradation of sulfamethoxazole and trimethoprim by ozone, ozone/hydrogen peroxide and ozone/persulfate processes: a comparative study. *Environmental Research* **189**, 109889.

Chen, M., Zhang, Z., Zhu, L., Wang, N. & Tang, H. 2019 Bisulfite-induced drastic enhancement of bisphenol A degradation in Fe^{3+}-H_2O_2 Fenton system. *Chemical Engineering Journal* **361**, 1190–1197.

Dong, C., Ji, J., Shen, B., Xing, M. & Zhang, J. 2018 Enhancement of H2O2 Decomposition by the Co-catalytic Effect of WS2 on the Fenton Reaction for the Synchronous Reduction of Cr(VI) and Remediation of Phenol. *Environmental Science & Technology* **52**, 11297–11308.

Feng, H., Liu, M., Zeng, W. & Chen, Y. 2020 Optimization of the O_3/H_2O_2 process with response surface methodology for pretreatment of mother liquor of gas field wastewater. *Frontiers of Environmental Science & Engineering* **15**, 78.

Feng, H., Liu, M., Zeng, W., Chen, Y., Wang, M., Yuan, L. & Yu, Z. 2022 Feasibility of resource utilization of the refractory evaporation concentrate of gas field wastewater exhibiting high salinity: Application of UV/Fenton, desulfurization, distillation and crystallization process after pre-treatment. *Environmental Research* **204**, 112317.

Fu, G. & Lee, J.-M. 2019 Ternary metal sulfides for electrocatalytic energy conversion. *Journal of Materials Chemistry A* **7** (16), 386–9405.

Garg, A., Singhania, T., Singh, A., Sharma, S., Rani, S., Neogy, A., Yadav, S. R., Sangal, V. K. & Garg, N. 2019 Photocatalytic degradation of bisphenol-A using N, Co codoped TiO_2 catalyst under solar light. *Scientific Reports* **9** (1), 765.

Guo, J., Zhou, Y., Yu, M., Liang, H. & Niu, J. 2021 The roles of wavelength in the gaseous toluene removal with OH from UV activated Fenton reagent. *Chemosphere* **275**, 129998.

Guo, W., Li, T., Chen, Q., Wan, J., Zhang, J., Wu, B. & Wang, Y. 2022 Construction of Fe^{2+}/Fe^{3+} cycle system at dual-defective carbon nitride interfaces for photogenerated electron utilization. *Separation and Purification Technology* **285**, 120357.

Hao, X., Zhou, J., Cui, Z., Wang, Y., Wang, Y. & Zou, Z. 2018 Zn-vacancy mediated electron-hole separation in ZnS/g-C_3N_4 heterojunction for efficient visible-light photocatalytic hydrogen production. *Applied Catalysis B: Environmental* **229**, 41–51.

Huang, C.-P. & Huang, Y.-H. 2008 Comparison of catalytic decomposition of hydrogen peroxide and catalytic degradation of phenol by immobilized iron oxides. *Applied Catalysis A: General* **346**, 140–148.

Ibrahim, M. F., Hod, R., Ahmad Tajudin, M. A. B., Wan Mahiyuddin, W. R., Mohammed Nawi, A. & Sahani, M. 2022 Children's exposure to air pollution in a natural gas industrial area and their risk of hospital admission for respiratory diseases. *Environmental Research* **210**, 112966.

Jiang, C., Pang, S., Ouyang, F., Ma, J. & Jiang, J. 2010 A new insight into Fenton and Fenton-like processes for water treatment. *Journal of Hazardous Materials* **174**, 813–817.

Jin, X., Zhang, L., Liu, M., Hu, S., Yao, Z., Liang, J., Wang, R., Xu, L., Shi, X., Bai, X., Jin, P. & Wang, X. C. 2022 Characteristics of dissolved ozone flotation for the enhanced treatment of bio-treated drilling wastewater from a gas field. *Chemosphere* **298**, 134290.

Kumar, N., Verma, S., Park, J., Chandra Srivastava, V. & Naushad, M. 2022 Evaluation of photocatalytic performances of PEG and PVP capped zinc sulfide nanoparticles towards organic environmental pollutant in presence of sunlight. *Chemosphere* **298**, 134281.

Lai, X., Ning, X.-a., Zhang, Y., Li, Y., Li, R., Chen, J. & Wu, S. 2021 Treatment of simulated textile sludge using the Fenton/Cl$^-$ system: the roles of chlorine radicals and superoxide anions on PAHs removal. *Environmental Research* **197**, 110997.

Lee, G.-J. & Wu, J. J. 2017 Recent developments in ZnS photocatalysts from synthesis to photocatalytic applications – a review. *Powder Technology* **318**, 8–22.

Li, Y., Dong, H., Li, L., Tang, L., Tian, R., Li, R., Chen, J., Xie, Q., Jin, Z., Xiao, J., Xiao, S. & Zeng, G. 2021 Recent advances in waste water treatment through transition metal sulfides-based advanced oxidation processes. *Water Research* **192**, 116850.

Liu, W., Wang, Y., Ai, Z. & Zhang, L. 2015 Hydrothermal synthesis of FeS$_2$ as a high-efficiency Fenton reagent to degrade alachlor via superoxide-mediated Fe(II)/Fe(III) cycle. *ACS Applied Materials & Interfaces* **7**, 28534–28544.

Liu, J., Dong, C., Deng, Y., Ji, J., Bao, S., Chen, C., Shen, B., Zhang, J. & Xing, M. 2018 Molybdenum sulfide co-catalytic Fenton reaction for rapid and efficient inactivation of *Escherichia coli*. *Water Research* **145**, 312–320.

Manny Porto Barros, M., Costa Almeida, K. J., Vinicius Sousa Conceição, M., Henrique Pereira & D., Botelho, G. 2023 Photodegradation of bisphenol A by ZnS combined with H$_2$O$_2$: evaluation of photocatalytic activity, reaction parameters, and DFT calculations. *Journal of Molecular Liquids* **371**, 121096.

Moussa, M. S., Sumanasekera, D. U., Ibrahim, S. H., Lubberding, H. J., Hooijmans, C. M., Gijzen, H. J. & van Loosdrecht, M. C. M. 2006 Long term effects of salt on activity, population structure and floc characteristics in enriched bacterial cultures of nitrifiers. *Water Research* **40**, 1377–1388.

Navalon, S., Miguel, M. D., Martin, R., Alvaro, M. & Garcia, H. 2011 Enhancement of the catalytic activity of supported gold nanoparticles for the Fenton reaction by light. *Journal of the American Chemical Society* **133**, 2218.

Osaka, T., Shirotani, K., Yoshie, S. & Tsuneda, S. 2008 Effects of carbon source on denitrification efficiency and microbial community structure in a saline wastewater treatment process. *Water Research* **42**, 3709–3718.

Rajeshwar, K., Osugi, M. E., Chanmanee, W., Chenthamarakshan, C. R., Zanoni, M. V. B., Kajitvichyanukul, P. & Krishnan-Ayer, R. 2008 Heterogeneous photocatalytic treatment of organic dyes in air and aqueous media. *Journal of Photochemistry and Photobiology C: Photochemistry Reviews* **9**, 171–192.

Ramirez, J. H., Maldonado-Hódar, F. J., Pérez-Cadenas, A. F., Moreno-Castilla, C., Costa, C. A. & Madeira, L. M. 2007 Azo-dye Orange II degradation by heterogeneous Fenton-like reaction using carbon-Fe catalysts. *Applied Catalysis B: Environmental* **75**, 312–323.

Ren, Y., Zhang, J., Ji, C., Wang, S., Lv, L. & Zhang, W. 2022 Iron-based metal-organic framework derived pyrolytic materials for effective Fenton-like catalysis: performance, mechanisms and practicability. *Science of The Total Environment* **809**, 152201.

Resources, M. o. N. 2021 *China Mineral Resources 2021*.

Sheng, F., Li, X., Li, Y., Afsar, N. U., Zhao, Z., Ge, L. & Xu, T. 2022 Cationic covalent organic framework membranes for efficient dye/salt separation. *Journal of Membrane Science* **644**, 120118.

Suave, J., Jose, H. J. & Fatima, P. 2018 Photocatalytic degradation of polyvinylpyrrolidone in aqueous solution using TiO$_2$/H$_2$O$_2$/UV system. *Environmental Technology* **39**, 1404–1412.

Wang, X., Wan, F., Han, K., Chai, C. & Jiang, K. 2008 Large-scale synthesis well-dispersed ZnS microspheres and their photoluminescence, photocatalysis properties. *Materials Characterization* **59**, 1765–1770.

Wang, M., Zhang, Q., Hao, W. & Sun, Z.-X. 2011 Surface stoichiometry of zinc sulfide and its effect on the adsorption behaviors of xanthate. *Chemistry Central Journal* **5**, 73.

Wang, Y., Li, H.-q. & Ren, L.-M. 2019 Organic matter removal from mother liquor of gas field wastewater by electro-Fenton process with the addition of H$_2$O$_2$: effect of initial pH. *Royal Society Open Science* **6**, 191304.

Wang, C., Zhao, Z., Deng, X., Chen, R., Liang, J., Shi, W. & Cui, F. 2021 Ultrafast oxidation of emerging contaminants by novel VUV/Fe^{2+}/PS process at wide pH range: performance and mechanism. *Chemical Engineering Journal* **426**, 131921.

Witte, B. D., Langenhove, H. V., Hemelsoet, K., Demeestere, K., Wispelaere, P. D., Van Speybroeck, V. & Dewulf, J. 2009 Levofloxacin ozonation in water: rate determining process parameters and reaction pathway elucidation. *Chemosphere* **76**, 683–689.

Xing, M., Xu, W., Dong, C., Bai, Y., Zeng, J., Zhou, Y., Zhang, J. & Yin, Y. 2018 Metal sulfides as excellent co-catalysts for H$_2$O$_2$ decomposition in advanced oxidation processes. *Chem* **4**, 1359–1372.

Xue, X., Hanna, K., Abdelmoula, M. & Deng, N. 2009a Adsorption and oxidation of PCP on the surface of magnetite: kinetic experiments and spectroscopic investigations. *Applied Catalysis B: Environmental* **89**, 432–440.

Xue, X., Hanna, K. & Deng, N. 2009b Fenton-like oxidation of rhodamine B in the presence of two types of iron (II, III) oxide. *Journal of Hazardous Materials* **166**, 407–414.

Zhang, Y. & Chu, W. 2022 g-C$_3$N$_4$ induced acceleration of Fe^{3+}/Fe^{2+} cycles for enhancing metronidazole degradation in Fe^{3+}/peroxymonosulfate process under visible light. *Chemosphere* **293**, 133611.

Zhang, J., Yu, J., Zhang, Y., Li, Q. & Gong, J. R. 2011 Visible light photocatalytic H-2-production activity of CuS/ZnS porous nanosheets based on photoinduced interfacial charge transfer. *Nano Letters* **11**, 4774–4779.

Zhu, B., Cheng, H., Ma, J., Kong, Y. & Komarneni, S. 2019 Efficient degradation of rhodamine B by magnetically separable $ZnS–ZnFe_2O_4$ composite with the synergistic effect from persulfate. *Chemosphere* **237**, 124547.
Zhu, B., Cheng, H., Qin, Y., Ma, J., Kong, Y. & Komarneni, S. 2020 Copper sulfide as an excellent co-catalyst with $K_2S_2O_8$ for dye decomposition in advanced oxidation process. *Separation and Purification Technology* **233**, 116057.
Ziembowicz, S. & Kida, M. 2022 Limitations and future directions of application of the Fenton-like process in micropollutants degradation in water and wastewater treatment: a critical review. *Chemosphere* **296**, 134041.

First received 19 November 2022; accepted in revised form 7 February 2023. Available online 22 March 2023

doi: 10.2166/wrd.2023.075

Oxidative stress, neurotoxicity, and intestinal microbial regulation after a chronic zinc exposure: an experimental study on adult zebrafish (*Danio rerio*)

Zixi Yuan[a], Ruixuan Li[a], Shuangshuang Li[b], Denggao Qiu[c], Guanyi Li[a], Chun Wang [a],*, Jiajia Ni[d], Yingxue Sun[a] and Hongying Hu[e]

[a] State Environmental Protection Key Laboratory of Food Chain Pollution Control, Key Laboratory of Cleaner Production and Integrated Resource Utilization of China National Light Industry, School of Ecology and Environment, Beijing Technology and Business University, Beijing 100048, China
[b] College of Energy and Environmental Engineering, Hebei University of Engineering, Handan 056038, China
[c] Key Laboratory of Cultivation and High-Value Utilization of Marine Organisms in Fujian Province, Fisheries Research Institute of Fujian, Xiamen 361000, China
[d] Research and Development Center, Guangdong Meilikang Bio-Science Ltd, Dongguan 523808, China
[e] Environmental Simulation and Pollution Control State Key Joint Laboratory, State Environmental Protection Key Laboratory of Microorganism Application and Risk Control (SMARC), School of Environment, Tsinghua University, Beijing 100084, China
*Corresponding author. E-mail: chun_wang@btbu.edu.cn

CW, 0000-0002-3475-2345

ABSTRACT

Zinc is one of the heavy metals present in textile wastewater with high concentrations. However, the chronic toxic effects of zinc on aquatic vertebrates are still ambiguous. Zinc accumulation in zebrafish after chronic zinc exposure and toxic effects on the intestines, muscles, and gills were investigated in this study. The results showed that a significant accumulation of zinc in the intestine, muscle, and gill was observed after 25 d of zinc exposure. The toxic effects of zinc were mainly in the form of zinc-induced oxidative stress in zebrafish, potential neurotoxicity, and changes in intestinal microbes. Significant changes in the levels of superoxide dismutase, catalase, metallothionein, glutathione, and malondialdehyde indicated that zinc damaged the antioxidant system of adult zebrafish. Zinc exposure resulted in a significant decrease in acetylcholinesterase activity and abnormal neural signaling. Furthermore, zinc exposure resulted in increased intestinal microbial richness and decreased the Simpson index in adult zebrafish. At the phylum and genus levels, the predominant microbes in the intestine are altered by zinc. In summary, this study provides an analysis of the toxic effects of chronic zinc exposure on adult zebrafish and the potential mechanisms, which are important for assessing the dual effects of zinc on aquatic organisms.

Key words: antioxidant, *Danio rerio*, intestinal microbiota, toxic effects, zinc

HIGHLIGHTS

- Zinc accumulation in adult zebrafish organs is significantly associated with oxidative stress.
- Differences in oxidative stress of different organs to chronic zinc exposure were found.
- Zinc adversely affects the nervous system of adult zebrafish.
- The effect of zinc on the intestinal microbiome of adult zebrafish is twofold.

<hr>

GRAPHICAL ABSTRACT

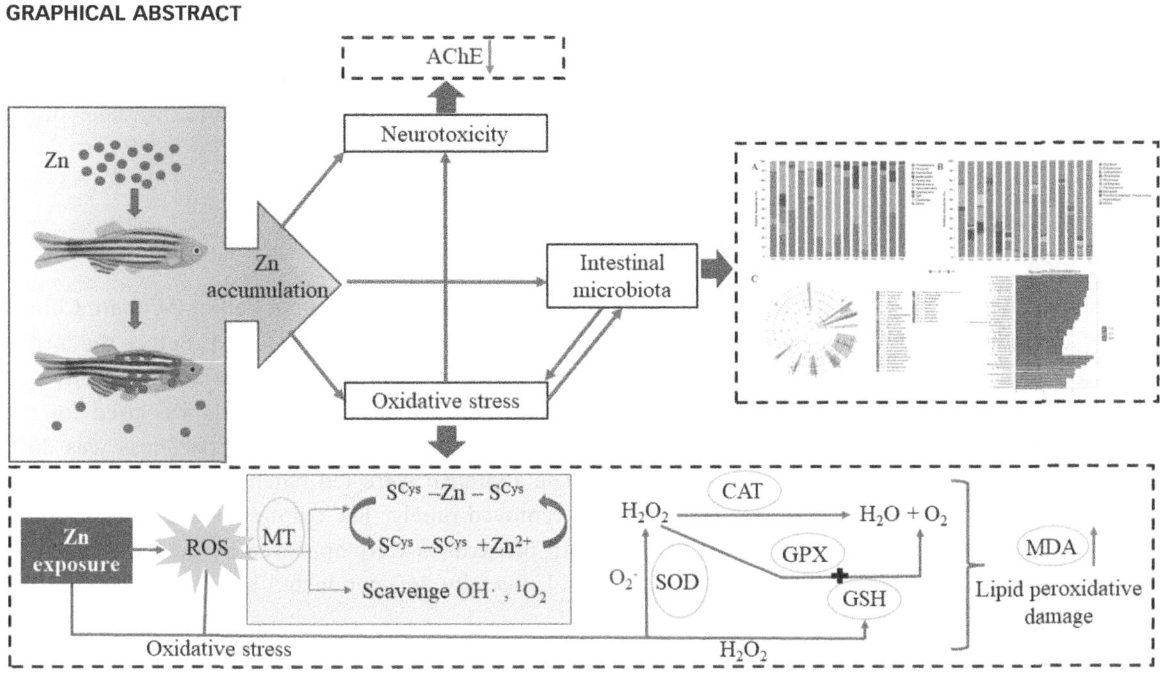

1. INTRODUCTION

The textile industry consumes a huge amount of water and is responsible for 17–20% of the total industrial pollution (Jegatheesan *et al.* 2016). Textile wastewater treatment plant effluent consists of a variety of complex chemicals such as acids, bases, salts, heavy metals, surfactants, oils, and fats. Especially, high levels of iron, copper, and zinc are included (El-Kassas & Mohamed 2014; Kishor *et al.* 2021; Nidheesh *et al.* 2022). Zinc is one of the heavy metals with a high ecological risk entropy in an aquatic environment (Supplementary Material, Figure S1). Elevated concentrations of zinc are also found in virtually all lakes and rivers in dense human-populated areas (Zheng *et al.* 2022). The high zinc concentrations in excess of 5 mg/L have been recorded in zinc-contaminated water environments, severely affecting the growth and development of aquatic life in rivers (Gozzard *et al.* 2011). However, most studies have focused on the toxic effects of zinc on fish embryos and larvae at high concentrations, while the mechanisms of chronic toxicity of zinc in adult zebrafish are still lacking.

Metalloenzymes and transcription factors contain zinc, which is essential for cellular growth, gene expression, protein synthesis, and cell division (Puar *et al.* 2021). However, zinc intake above nutritional levels poses a risk of toxicity to fish. It has been reported that zinc exposure at 1.5 and 4.9 mg/L resulted in delayed hatching in zebrafish and with the increase of zinc concentration, zebrafish mortality increased (Horie *et al.* 2020). Disruption of Ca, Mn, and Co homeostasis in zebrafish larvae by zinc exposure has also been observed (Puar *et al.* 2021). In addition, an investigation demonstrated that *Etroplus suratensis* developed significant necrotic lesions in the gills under 15.32 mg/L zinc exposure (Xavier *et al.* 2019).

The intestines, gills, and muscles are all critical organs for aquatic vertebrates. In addition to nutrient absorption and metabolism, the intestine also regulates the intrinsic immune system and maintains metal homeostasis (Zeng *et al.* 2019). Fish intestine can be used to evaluate toxicological effects when heavy metal contamination occurs (Zeng *et al.* 2019). Heavy metals mainly cause fish intestinal histopathological lesions (Dane & Sman 2020), imbalance of intestinal microbial community, damage of antioxidant enzymes and neural toxicity (Wang *et al.* 2020). Gills are involved in gaseous and ionic exchanges between fish and the water environment, and are easily affected by aqueous pollutants of their large surface area and small diffusion distance (Santos *et al.* 2022). Heavy metal exposure causes dilation of blood vessels in fish gills, affecting the blood supply to other organs (Luzio *et al.* 2021). Muscle is the largest tissue of fish and is the main effector of fish swimming behavior (Shahjahan *et al.* 2022). Damage to muscle cells can cause behavioral changes emerg abnormal feeding ability in zebrafish (Avallone *et al.* 2015). Therefore, abnormal changes in zebrafish intestines, muscles, and gills can be used as evidence to demonstrate the toxic effects of excessive zinc exposure in aquatic vertebrates.

To determine the mechanism of zinc accumulation and biotoxic effects in the intestines, muscles, and gills of fish following chronic zinc exposure, oxidative damage and neurotoxicity effects of chronic zine exposure in zebrafish organs were analyzed. Meanwhile, high-throughput sequencing was used to evaluate the effect of zinc on the zebrafish intestinal microbiome. By comparing the effects of different concentrations of zinc on zebrafish, it provides a new idea for dialectically regard with the harm of excessive zinc.

2. MATERIALS AND METHODS

2.1. Zebrafish maintenance

Adult wild-type zebrafish (AB, 3–4 months old) were purchased from the China Zebrafish Resource Center (Wuhan, China). Zebrafish were 3.0 ± 0.5 cm in body length and 0.4 ± 0.05 g in weight. The male-to-female ratio was 1:1. Before the formal experiment, zebrafish were uniformly acclimated in a large fish aquarium (65 L) for 7 d to adapt to the laboratory environment and ensure a natural mortality rate below 5%. Dechlorinated tap water, well aerated for 7 d, was used for the zebrafish culture. During the experiment, the water temperature was $25 \pm 2\,°C$, pH was 7.7 ± 0.1, hardness was 200–230 mg/L $CaCO_3$, and a 12 h/12 h light–dark cycle was maintained. The zebrafish were fed hatched brine shrimp twice daily at 9:00 and 18:00. Dead fish, excrement, and food residues were removed timely. The culture solution was changed every 2 d to ensure a constant concentration of zinc in the fish tank. All procedures were approved by the Committee on the Ethics of Animal Experiments of Beijing Technology and Business University and conducted in accordance with the guidelines for the protection of animal welfare.

2.2. Experimental design and sample collection

$ZnCl_2$ (CAS No. 7646-85-7, purity: 99.95%) was purchased from Beijing Mreda Technology Co., Ltd (Beijing, China) and used to regulate zinc concentration in culture water. Based on previous pre-experimental results (Supplementary Material, Figure S2) and environmental investigation data (Gozzard *et al.* 2011), the zinc concentration in the formal experiment was set to CG (control group), LG (low concentration group), and HG (high concentration group) with 0, 5, and 10 mg/L of zinc concentrations, respectively. Three replicate experiments were conducted for each concentration. After 7 d of acclimation, 270 adult zebrafish were equally divided into nine tanks (5 L) for a 25-d chronic exposure experiment. Two zebrafish were randomly collected from each tank and anesthetized with tricaine (MS-222; Aladdin Biochemical Technology Co., Ltd, Shanghai, China). After dissection, the intestines, muscles, and gills of fish were sampled for analyzing zinc levels and antioxidant enzyme assays. In addition, two fish samples were randomly collected from each tank and anesthetized with MS-222, and then only intestinal samples were taken for 16S rRNA sequencing. Samples were immediately frozen at $-80\,°C$ for subsequent analysis.

2.3. Measurement of biochemical indicators

The zinc content in different zebrafish tissues was measured at 5 and 25 d. The tissue was accurately weighed and the volume of deionized water was added nine times to make a 10% tissue homogenate. It was centrifuged at 2,500 rpm for 10 min, and 25 μL of the supernatant was taken to measure the zinc ion concentration. Antioxidant index levels and acetylcholinesterase (AChE) activity were measured at exposure times of 5, 10, 15, 20, and 25 d, respectively. Zebrafish intestine, muscle, and gill tissue samples were weighed accurately, and pre-cooled saline was added at a ratio of weight (g):volume (mL) = 1:10. The samples were processed using a high-speed grinder, and 10% of the tissue homogenates were prepared. Samples were then centrifuged at 2,500 rpm for 10 min. After dilution, the supernatant was collected. Antioxidant enzymes (superoxide dismutase (SOD), catalase (CAT) and glutathione (GSH)), malondialdehyde (MDA), and metallothionein (MT) were tested using assay kits (Nanjing Jiancheng Bioengineering Institute, Jiangsu, China) for indicating oxidative stress. In this study, AChE was analyzed using an assay kit (Nanjing Jiancheng Bioengineering Institute) as a recognized marker of neurotoxicity.

2.4. 16S rRNA sequencing and bioinformatics analysis

For intestinal microbiota composition analysis, the total DNA of the intestinal samples was extracted using the DNeasy Blood & Tissue Kit (Qiagen GmbH, Hilden, Germany; reference number 69506). The V3–V4 region of the bacterial 16S rRNA gene was amplified using primers 338F (5′-ACTCCTACGGGAGGCAGCA-3′) and 806R (5′-GGACTACHVGG GTWTCTAAT-3′) as previously described (Castrillo *et al.* 2017). After purification, all amplicons were pooled together with an equal molar amount from each sample and sequenced using an Illumina platform. The high-throughput sequencing of intestinal microbiome samples was all completed by Shanghai Personalbio Technology Co., Ltd.

Vsearch (v2.13.4_linux_x86_64) and cutadapt (v2.3) were used for further analysis of high-throughput sequencing data. After de-priming, splicing, quality filtering, and de-chimera, high-quality sequences were clustered at a sequence similarity threshold of 97%, and the representative sequence and OTU tables were output. Representative sequences of OTUs were annotated by species using the classify-sklearn algorithm of QIIME2 (Bokulich *et al.* 2018). Finally, differences in species diversity and community structure between samples were revealed through α- and β-diversity analyses (Ding *et al.* 2019). Principal coordinate analyses (PCoAs) were performed at the OTU level using the R 3.3.1 (The R Foundation for Statistical Computing, Vienna, Austria). A PCoA plot was obtained based on the Bray–Curtis distances between the samples. Rarefaction curves of sample abundance were plotted, and α-diversity was evaluated using the Chao1 richness and the Simpson diversity index.

2.5. Statistical analysis

Mean and standard error of the mean (SEM) are used to present all data for each group. To determine the significant differences between oxidative stress marker levels, zinc content, and AChE activity between groups, SPSS software was used to conduct a one-way analysis of variance (ANOVA) followed by the least significant difference (LSD) test (SPSS 25.0, IBM, Armonk, New York, NY, USA). Graphs were constructed using Origin 2019 (Microcal, Redmond, Washington, USA). The significance of differences in microbial communities was verified by the Kruskal–Wallis *post hoc* test (XL-Stat software, Addinsoft). Differences were considered statistically significant at $p < 0.05$.

3. RESULTS

3.1. Changes of zinc content in zebrafish tissues

Cumulative mortality in zebrafish is less than 30% in chronic zinc exposure experiments (Supplementary Material, Figure S3). Significantly higher zinc levels were observed in the intestines, muscles, and gills of the experimental groups than those in the control group ($p < 0.05$; Figure 1). In the case of the intestines of the 5 mg/L group, the zinc content at 25 d was approximately threefold higher than that at 5 d. Furthermore, zinc levels in the intestines, muscles, and gills were significantly higher at 25 d than 5 d ($p < 0.05$; Figure 1 and Supplementary Material, Figure S4).

3.2. Effect of zinc exposure on the antioxidant system of zebrafish

3.2.1. MT level

The intestinal MT of zebrafish exposed to 10 mg/L zinc was significantly higher than that of the control group ($p < 0.05$). In the first 10 d, the MT levels in the 5 mg/L group were not significantly different from those of the control group. In the 5 mg/L group, MT at 15–25 d showed significant fluctuation and reached the highest level at 15 d (Figure 2(a)). No significant difference in MT in the muscles of each experimental group was found at 5–10 d (Figure 2(b)). After 15 d, the MT level in the muscle of zebrafish in the 5 mg/L group decreased with time and was significantly lower than that in the control group. A

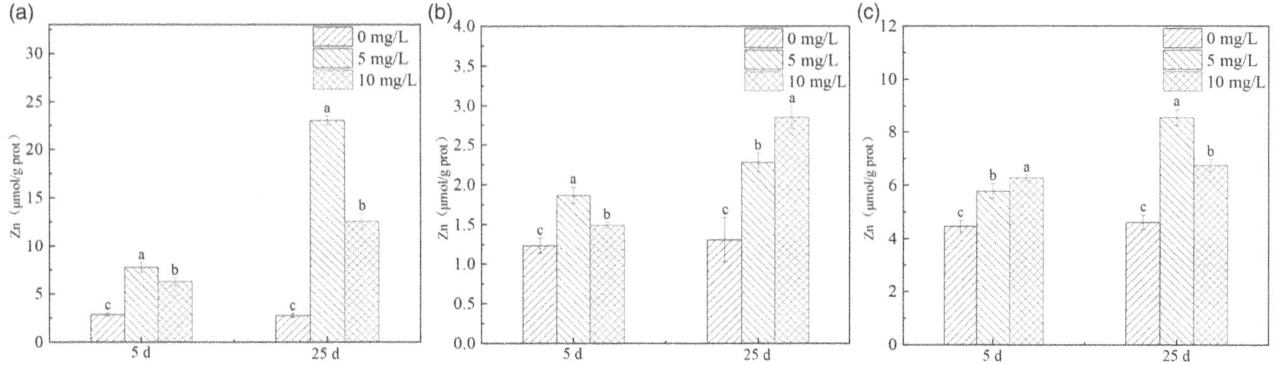

Figure 1 | Zinc accumulation levels in zebrafish tissues. (a) Intestine; (b) muscle; and (c) gill. Columns with different letters indicated values with significant difference ($p < 0.05$).

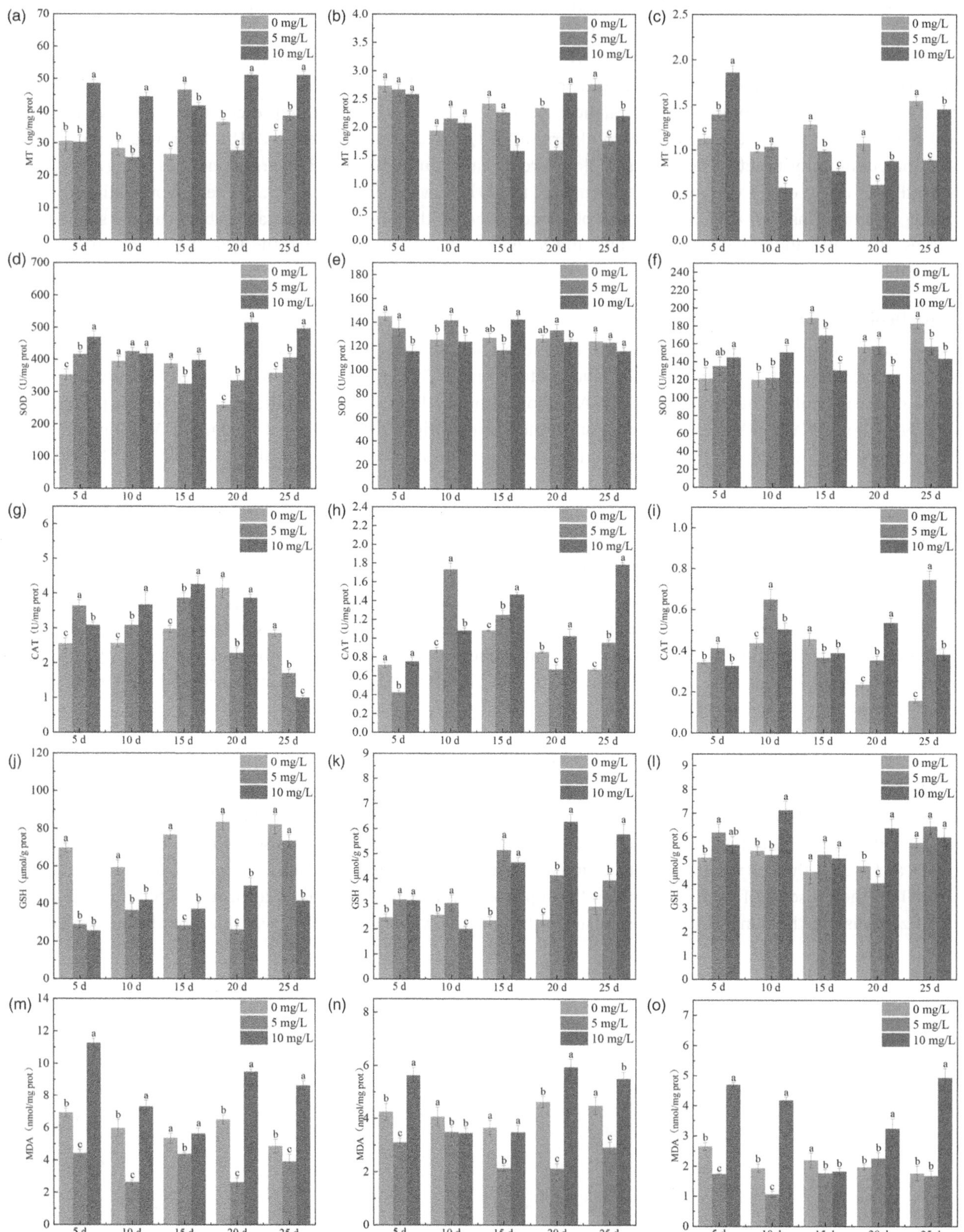

Figure 2 | Activity levels of MT, SOD, CAT, GSH, and MDA content in the organs of zebrafish during zinc exposure. (a), (d), (g), (j), and (m) Intestine; (b), (e), (h), (k), and (n) muscle; (c), (f), (i), (l), and (o) gill. Data are shown as mean ± SD ($n = 5$). Columns with different letters indicated values with significant difference ($p < 0.05$).

fluctuating trend of MT in the 10 mg/L group was observed (Figure 2(b)). Significantly higher MT levels were found in the experimental groups at 5 d and significantly decreased with longer exposure time (>15 d; Figure 2(c)).

3.2.2. SOD activity

Intestinal SOD activity of adult zebrafish exposed to 5 and 10 mg/L zinc was significantly higher than that of the control group at 20–25 d ($p < 0.05$; Figure 2(d)). However, no significant differences in intestinal SOD activity were observed between the control and exposure groups after 10–15 d. SOD activity in the zebrafish muscle was maintained at a similar level (Figure 2(e)). Although there was no significant difference in SOD activity in zebrafish gills among the three groups at 5 d, from day 10, a significant difference was observed between the experimental and control groups ($p < 0.05$; Figure 2(f)).

3.2.3. CAT activity

Exposure to different concentrations of zinc had significant effects on zebrafish intestinal CAT activity (Figure 2). Specifically, from days 5 to 15, intestinal CAT activities in the experimental groups were significantly higher than those in the control group ($p < 0.05$). However, opposite results were found at 20 and 25 d (Figure 2). Furthermore, CAT activity in the 10 mg/L experimental group showed a trend of first increasing and then decreasing over time (Figure 2). Short-term (5 d) zinc exposure did not significantly impact CAT activities in zebrafish muscles and gills (Figure 2 and 2(i)). At 25 d, a significant increase in CAT activity in muscles and gills was found between the experimental and control groups (Figure 2 and 2(i)).

3.2.4. GSH content

The GSH contents in the intestine of zebrafish in the experimental groups were significantly lower than those in the control group ($p < 0.05$; Figure 2(j)). The effects of zinc exposure on intestinal GSH content in zebrafish were similar between the 5 and 10 mg/L groups during the initial 10 d, but significant differences were observed from day 15 onwards ($p < 0.05$). The GSH levels in the muscles of the experimental groups were significantly higher than those of the control group at all sampling times except the 10 mg/L group on 10 d (Figure 2k). GSH content in the gills of zebrafish was not statistically significant in the experimental groups compared with that in the control group ($p > 0.05$; Figure 2l).

3.2.5. MDA content

MDA content in the zebrafish intestine decreased significantly in the 5 mg/L treatment group compared to that in the control, whereas a significant increase in MDA was found in the 10 mg/L group ($p < 0.05$; Figure 2m). The trends in MDA content in zebrafish muscle were similar to those in the intestine (Figure 2(n)). In contrast, no significant difference in MDA content was observed between the 5 and 10 mg/L groups at 10 d ($p > 0.05$; Figure 2(n)). The MDA content was significantly increased in zebrafish gills after exposure to 10 mg/L zinc (Figure 2(o)). Moreover, MDA content decreased significantly in the 5 mg/L treatment group at days 5–15 of exposure compared to that in the control ($p < 0.05$), and no significant change was observed at days 20 and 25 ($p > 0.05$; Figure 2(o)).

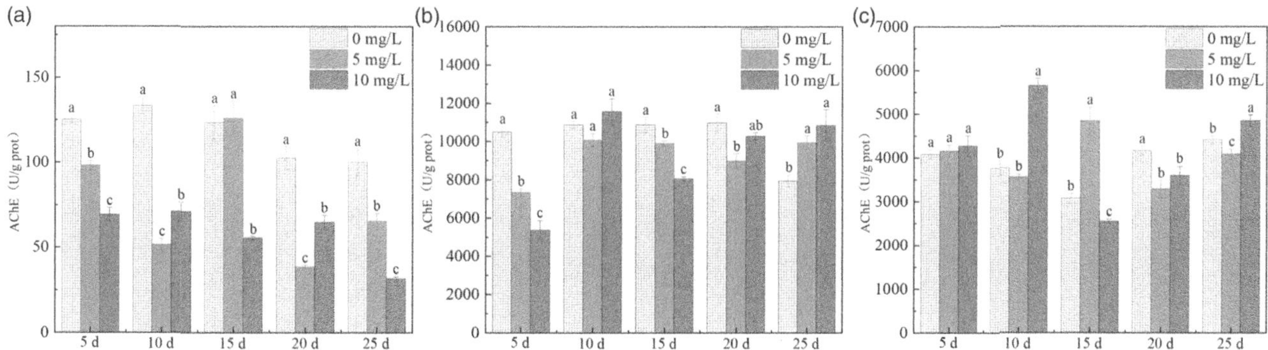

Figure 3 | Activity levels of AChE in the organs of zebrafish during zinc exposure. (a) Intestine; (b) muscle; and (c) gill. Columns with different letters indicated values with significant difference ($p < 0.05$).

3.3. Effects of zinc exposure on AChE in zebrafish

Zinc exposure resulted in a significant decrease in AChE activity in the intestine and muscle compared with the controls ($p <$ 0.05; Figure 3). However, AChE activity in the muscle significantly increased after 25 d of zinc exposure (Figure 3(a) and 3(b)). At day 5, there was no significant difference in AChE activity in the gills between the experimental and control groups. On days 10–25 of the experiment, inconsistently significant changes were observed in the AChE activity of the experimental groups ($p < 0.05$; Figure 3(c)). AChE activity and physiological indicators of oxidative stress were correlated (Table 1), and AChE was significantly correlated with SOD, GSH, and MT at 95% confidence intervals ($p < 0.01$).

3.4. Differences in the levels of detectable indicators in different tissues

The levels of SOD, GSH, and MT in the intestinal tract of the experimental groups were higher than those in the muscles and gills (Figure 2), which is consistent with the accumulation of zinc (Figure 1). Correlation analysis (Table 2) demonstrated a highly significant correlation between zinc accumulation in organs and changes in SOD, GSH, MT, and AChE levels ($p < 0.01$). MDA and CAT levels in intestines, muscles and gills were not significantly correlated with zinc levels (Table 1). The differences in the distribution of MDA in the intestines, muscles, and gills were not significant. In addition, the activity of AChE in tissues was in the order of muscle > gill > intestine (Figure 3), but AChE was most significantly affected by zinc in the intestine ($p < 0.05$; Figure 3(a)).

3.5. Intestinal microbiota structure analysis

In total, 7,006 OTUs were obtained with a similarity level of 97%. A plateau shape was observed for the Chao1 rarefaction curve of zebrafish intestinal microbiota constructed using OTU clustering results (Supplementary Material, Figure S5). Based

Table 1 | Correlation of AChE activity with other parameters

AChE						95% Confidence interval	
	Pearson correlation coefficient	Sig. (two-tailed)	Number of cases	Deviation	Standard error	Lower limit	Upper limit
CAT	−0.217	0.277	27	−0.006	0.190	−0.561	0.207
SOD	−0.858**	0.000	27	−0.005	0.029	−0.911	−0.798
GSH	−0.781**	0.000	27	−0.005	0.041	−0.862	−0.701
MT	−0.790**	0.000	27	−0.005	0.047	−0.873	−0.682
MDA	−0.081	0.687	27	0.008	0.181	−0.400	0.302
AChE	1		27	0	0	1	1

**$p < 0.01$. Indicates significant correlation at the 0.01 level (two-tailed).

Table 2 | Correlation of zinc accumulation levels with other parameters

Zn						95% Confidence interval	
	Pearson correlation coefficient	Sig. (two-tailed)	Number of cases	Deviation	Standard error	Lower limit	Upper limit
Zn	1		27	0	0	1	1
CAT	0.216	0.280	27	−0.007	0.163	−0.163	0.488
SOD	0.660**	0.000	27	0.002	0.098	0.440	0.831
GSH	0.524**	0.005	27	−0.008	0.203	0.098	0.922
MT	0.627**	0.000	27	−0.001	0.107	0.373	0.807
MDA	−0.005	0.982	27	−0.002	0.167	−0.322	0.380
AChE	−0.639**	0.000	27	−0.003	0.104	−0.829	−0.422

**$p < 0.01$. Indicates significant correlation at the 0.01 level (two-tailed).

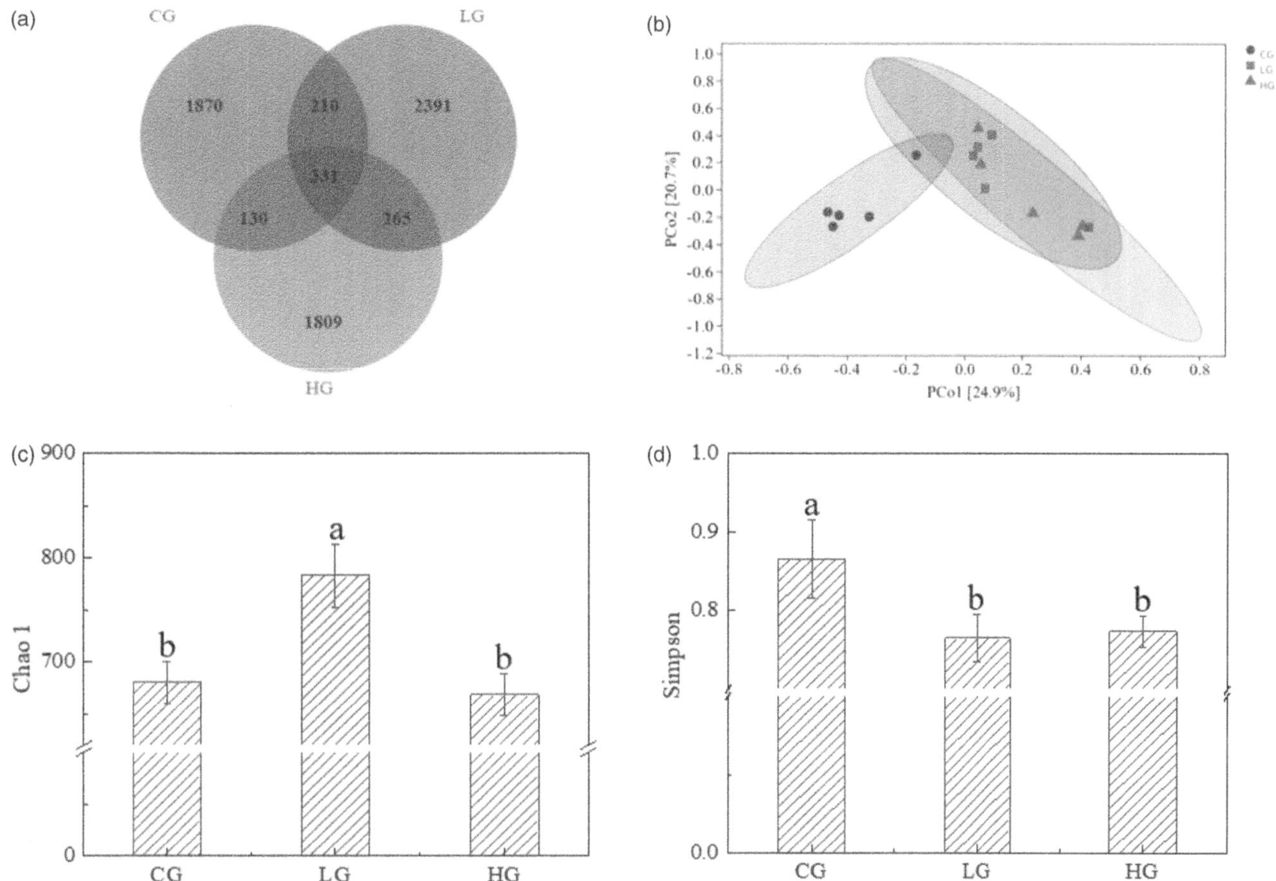

Figure 4 | Effects of different treatments on the community structure of intestine microbiota in zebrafish. (a) Venn diagram showing the number of unique and shared OTUs in different treatments. (b) PCoA revealing the differences in the community structure between different treatments. Asterisks indicated significant differences. (c) Chao1 and (d) Simpson indexes showing the diversity of intestine microbiota in different treatments. Columns with different letters indicated values with significant difference ($p < 0.05$). CG, LG, and HG represent experimental groups with zinc concentrations of 0, 5, and 10 mg/L, respectively.

on the sequencing data, it could be assumed that most of the microbial diversity could be covered, so the vast majority of microbial information could be analyzed. Chronic exposure of zebrafish to 0, 5, and 10 mg/L zinc produced 2,541, 3,197, and 2,535 OTUs, respectively. All groups shared 331 OTUs (Figure 4(a)).

In this study, to determine the microbiota α-diversity of zebrafish, the Chao1 and Simpson indices were used. The Chao1 index significantly increased in the 5 mg/L (LG) treatment group ($p < 0.05$; Figure 4(c)). Significantly lower Simpson indexes were found in both LG and HG than in the control group ($p < 0.05$; Figure 4(d)). In addition, there was no significant difference in the Simpson index between the LG and HG groups ($p > 0.05$; Figure 4(d)). The results of the PCoA based on the Bray–Curtis distance were consistent with those of the Simpson index. According to the PCoA, the 5 and 10 mg/L zinc exposure groups clearly deviated from the control group (Figure 4(b)). The zebrafish intestinal microbiota has been shown to change after chronic exposure to zinc. The LG and HG groups did not completely overlap, demonstrating that different concentrations of zinc produced different effects on intestinal microbiota.

In the zebrafish intestine, the major phyla were Proteobacteria, Firmicutes, Fusobacteria, Bacteroidetes and Tenericutes (Figure 5(a)). A large proportion (95%) of the relative abundance in the control group was attributed to these bacterial communities (Supplementary Material, Figure S6). The relative abundances of Fusobacteria and Tenericutes were lower in the experimental groups than in the control group. Fusobacteria in the LG and HG groups decreased by 29.34 and 29.33%, respectively, compared to that in the control group (CG: 29.41%, LG: 0.07%, and HG: 0.08%), and the relative abundance of Tenericutes decreased by approximately 15.5% (CG: 15.5%, LG and HG < 0.01%; Supplementary Material, Table S1).

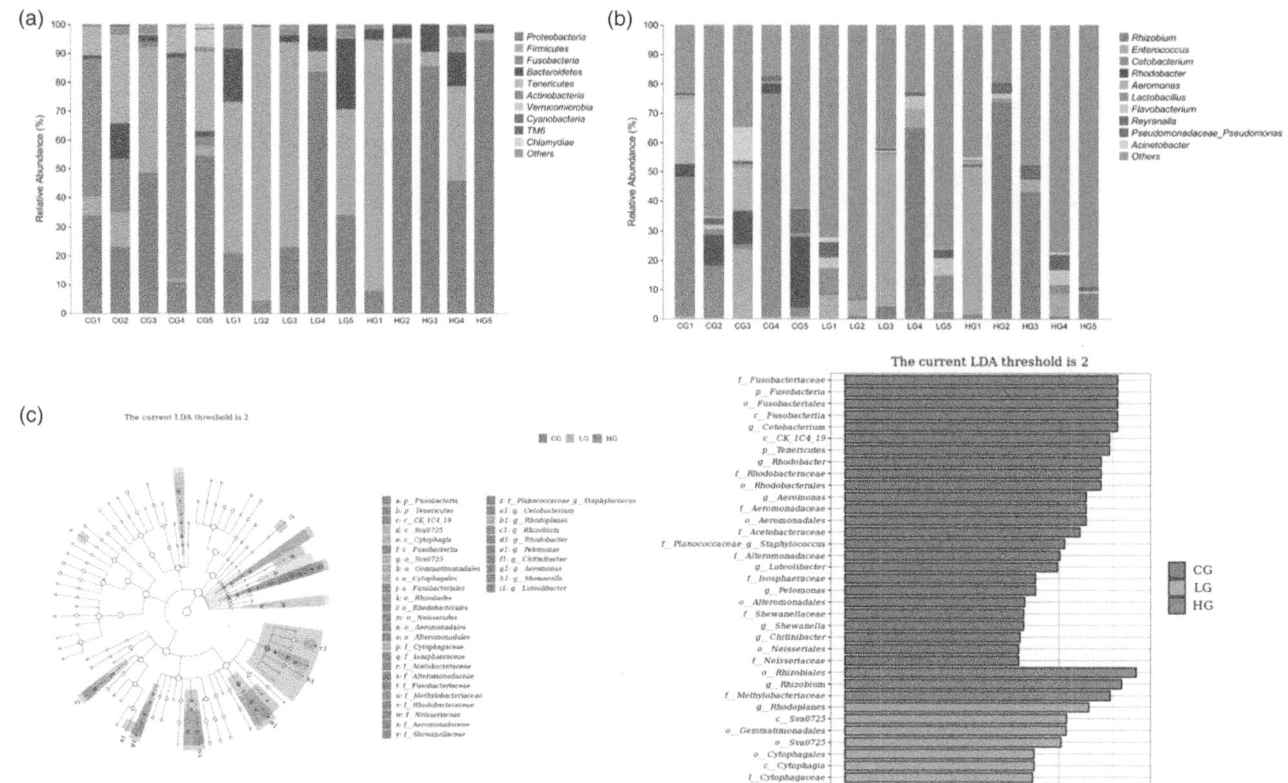

Figure 5 | Bacterial composition of the various communities at the phylum (a) and genus levels (b) Taxa with abundances <1% are included in 'others'. (c) LEfSe taxonomic cladogram. The colored nodes from inner to outer circles represent the hierarchical relationship of all taxa from the phylum to the genus level. Blue, red, and green circles indicate differences in the relative abundance; white circles indicate non-significant differences. Please refer to the online version of this paper to see this figure in colour: http://dx.doi.org/10.2166/wrd.2023.075.

Zinc exposure increased the relative abundance of Proteobacteria, Bacteroidetes, and Firmicutes. Bacteroidetes in the experimental groups increased by 6.2 and 2.08% in the LG and HG groups, respectively (CG, 3.87%; LG, 10.07%; and HG, 5.95%; Supplementary Material, Table S1). The experimental groups showed a concentration-dependent relationship. An increase in Firmicute abundance was observed in the 5 mg/L group (LG). However, the effect of increased Proteobacteria abundance was more pronounced in the 10 mg/L group (HG). The relative abundance of Firmicutes increased by 39.01% in the LG group (CG, 13.49%; and LG, 52.50%) and by 31.27% in the HG group (CG, 34.05%; and HG, 65.32%; Supplementary Material, Table S1).

At the genus level, *Cetobacterium* and *Rhodobacter* were the most abundant in the control group. Exposure to zinc in the experimental groups significantly decreased the relative abundance of *Cetobacterium* and *Rhodobacter*, which may be related to the occurrence of diseases (Figure 5(b)). In contrast, zinc exposure promoted an increase in the relative abundance of *Rhizobium* and *Enterococcus*. These changes were confirmed by linear discriminant analysis (LDA) and effect size (LEfSe) analysis (LDA ≥ 2, Figure 5(c)). Through LEfSe analysis, 35 distinct taxa were detected in the intestinal microbiota of the treatment groups compared with the control group. Additionally, the relative abundances of several other species of bacteria involved in vital physiological and biochemical processes changed significantly. For example, *Aeromonadales*, *Gemmatimonadetes*, and *Rhizobiales* had very high LDA scores (more than four orders of magnitude) in these samples (Figure 5(a) and 5(b)). It is noteworthy that the higher the LDA score, the greater the effect of this taxon on the differences between the groups. Heatmap based on dominant genus compositions of intestinal microbiota showed that the experimental groups were significantly different from the control group, and chronic exposure to zinc significantly altered the microbial community structure of the adult zebrafish intestine (Supplementary Material, Figure S7).

4. DISCUSSION

The process of transmission and accumulation of heavy metals in the gills, intestines, and muscles of fish is described. Fish gills are used for gas exchange in aquatic environments. Large gill surface areas with abundant active ion transport pumps are capable of efficiently absorbing heavy metals (Zhang *et al.* 2022). The absorbed zinc is exchanged with blood under the action of the gills, which promotes the distribution of zinc to the intestines and muscles (Juncos *et al.* 2019). This process prevents the excessive accumulation of zinc in the gills (Tsai & Liao 2006). The present study showed that after 25 d of zinc exposure, there was a significant accumulation of zinc in the organs of zebrafish, and the distribution of accumulation was in the order of intestine > gill > muscle. It has been reported that the accumulation level of cadmium in fish organs is in the order of intestine > kidney > liver > gill > muscle (Wang *et al.* 2020). This report can be cross-checked with the results of our study. In addition, the accumulation of toxic substances in the zebrafish intestine was found to be higher than that in the gills and muscles (Zhang *et al.* 2022).

The cumulative mortality of adult zebrafish approaching 30% during chronic experiments is the result of multiple mechanisms acting together in an environment of high zinc stress. The main causes of mortality are likely to be (1) excessive oxidative damage triggered by high zinc concentrations leading to the collapse of the antioxidant and immune systems, triggering death by immune disease (Capriello *et al.* 2021). (2) Dysbiosis of the intestinal microflora leads to disruption of the metabolic system and abnormal signaling of the nervous system, which seriously affects the normal life activities of zebrafish (McRae *et al.* 2016). (3) Excessive accumulation of Zn disturbs metal homeostasis, such as disrupting the calcium absorption mechanism, and eventually the fish suffer from hypocalcemia and die (Xu *et al.* 2019). (4) Excessive Zn transport through the gills causes branchial mucus secretion, which has a damaging effect on the gills, thus limiting the transport and absorption of oxygen by the gills (Skidmore 1970).

Zinc is a redox-inert divalent metal ion involved in the regulation of redox processes through interactions with cysteine sulfur in cellular proteins (Maret 2019). The cysteine-rich protein in cells is MT, which binds up to seven zinc ions in the fully reduced state (Hubner & Haase 2021). MT is used for the detoxification of heavy metals, such as zinc, and for antioxidant regulation *in vivo*. The main regulatory process involves zinc binding to reduced thiols and their release under oxidative conditions. The redox ligand formed by zinc as the central ion converts the redox signal to a zinc ion concentration (zinc signal) while participating in the redox process. Zinc signals can trigger an antioxidant response (Maret 2006). In addition, MT can directly scavenge harmful hydroxyl radicals and singlet oxygen to cooperate with oxidative detoxification (Gimenez *et al.* 2021).

The MT of the 10 mg/L group in the zebrafish intestine was maintained at a high level. This reflects the increasing damage to the intestinal tract when zinc levels are high. A large amount of MT is generated to scavenge the reactive oxygen species (ROS). At the early stage of zinc exposure (5 d) in the 5 mg/L group, MT in the gills increased most significantly, because the gills are the channel for ingesting zinc. MT is increased to bind more zinc ions for detoxification while scavenging hydroxyl and oxygen radicals more effectively. With prolonged exposure, the gills transfer the accumulated zinc ions to organs, such as the intestine, resulting in a decrease in MT in the gills and a significant increase in MT in the intestine. Consistent with the conclusions of the present study, zinc exposure has been reported to cause an increase in MT levels in the gills of zebrafish, followed by a decrease (Arini *et al.* 2015).

Excess zinc activates lipoamide dehydrogenase in zebrafish cells, which, in turn, catalyzes the massive production of hydrogen peroxide and superoxide radicals (Lee 2018). At this time, the balance between the production and elimination of ROS is disrupted, resulting in oxidative stress. Biological defenses against oxidative stress have evolved in fish (Sun *et al.* 2020). Antioxidative enzymes play crucial roles in cellular antioxidant processes. SOD converts highly reactive superoxide radicals into water and hydrogen peroxide, which are further decomposed into non-toxic oxygen and water by CAT (Shahjahan *et al.* 2022). In addition, the cysteine group contained in the reduction of GSH by non-enzymatic compounds has redox activity and participates in the eradication of hydrogen peroxide, together with antioxidant enzymes (Wu *et al.* 2019). However, the regulatory capacity of the antioxidant systems is limited. When the regulatory capacity of the antioxidant enzyme system is exhausted, damage owing to cellular lipid peroxidation cannot be avoided (Sun *et al.* 2020). One of the best indicators of oxidative damage is MDA, which is the final product of lipid peroxidation (Sun *et al.* 2019).

The present study showed that short-term (5 d) zinc exposure significantly increased SOD activity in the intestine, which is a marker for the activation of the antioxidant system. Zebrafish are stressed by zinc, and excessive ROS production in cells stimulates SOD activity to effectively regulate the balance of ROS (Sharma *et al.* 2022). In the mid-exposure period (10–15 d), under the action of SOD, the level of ROS gradually returned to a steady state, and SOD returned to its original state. However, SOD was reactivated with the excessive accumulation of zinc with prolonged exposure time. Muscles and gills also showed a decreased level of SOD activity. Enzyme activity is impaired or their generation is declining due to tissue damage, as evidenced by the decline in enzyme activity (Sharma & Jindal 2020). Significant changes in SOD activity after exposure to environmental pollutants have been reported (Jiang *et al.* 2022).

CAT activity in the intestines, muscles, and gills of zebrafish showed an overall trend of enhancement during zinc exposure, which was similar to that of SOD activity. This suggests a synergistic effect of SOD and CAT on ROS scavenging (Si *et al.* 2019). Increased CAT activity indicates activation of the cellular antioxidant defense system to repair oxidative stress damage by scavenging and breaking down excess H_2O_2 (Sharma *et al.* 2022). Specifically, intestinal CAT activity decreased significantly after 20–25 d. This may be because of the large amount of CAT used to eradicate H_2O_2 after a sustained H_2O_2 surge after continuous zinc exposure. When CAT production was slower than consumption, the antioxidant system was close to collapse. Thus, the conclusion that CAT activity is activated at low toxicity and inhibited at high toxicity was validated (Jiang *et al.* 2022).

A plausible explanation for the significant reduction of intestinal GSH of zebrafish in the experimental groups is that GSH peroxidase catalyzes the reduction of H_2O_2 at the expense of GSH (Capriello *et al.* 2021). Moreover, glutathione S-transferase induces the binding of electrophilic ions to GSH to form conjugates to maintain cellular redox homeostasis (Wu *et al.* 2019). GSH levels in the experimental groups were lower than those in the control group; this phenomenon is common in zebrafish studies. Oxidative stress responses induced by metals such as cadmium (Hu *et al.* 2022) and aluminum (Capriello *et al.* 2021) have also been observed. The present data showed that GSH levels were increased in zebrafish muscle. We speculate that this may be owing to the fact that muscle is less affected by oxidative stress, which is an initial response to ROS following the accumulation of trace toxicity of zinc.

The decreased MDA levels in the low concentration group implied that the defense effect of the antioxidant system was effectively exerted in the 5 mg/L zinc exposure group. The cells were not damaged by lipid peroxidation. Conversely, high levels of zinc induced significant ROS production. Invasion by ROS initiates antioxidant defense activities in the intestines, muscles, and gills of zebrafish. However, the defense system is insufficient for complete removal, leading to lipid peroxidation (Sharma *et al.* 2022). This was confirmed by the significant increase in MDA levels in the 10 mg/L experimental group. In a previous report on the effects of Cd on the liver of zebrafish, no significant difference in MDA levels was found at low concentrations, and high concentrations of Cd led to a significant increase in MDA (Hu *et al.* 2022).

AChE plays an important role in nerve impulse conduction and is a recognized biomarker of neurotoxicity in toxicology studies (Saiki *et al.* 2021). The enzyme immediately terminates the continuous excitatory effect of neurotransmitters by hydrolyzing acetylcholine, ensuring the normal transmission of nerve signals in the body (Tao *et al.* 2022). Zinc exposure significantly inhibits AChE activity in the intestines, muscles, and gills of zebrafish as was observed in the present study. This may be because of oxidative stress (Muthulakshmi *et al.* 2018). Furthermore, oxidative stress may alter AChE activity, and this finding is supported by other studies (Parlak 2018; Pullaguri *et al.* 2020). A significant enhancement in AChE activity in the muscle was observed after 25 d of exposure. A reasonable hypothesis is that lipid peroxidation damage leads to the rupture of presynaptic vesicle membranes and a large amount of acetylcholine is released, inducing a significant enhancement of AChE activity. Similar phenomena have been reported in another study (Zhang *et al.* 2021).

In the analysis of oxidative stress, the levels of SOD, CAT, GSH, and MT in the intestine of the experimental groups were higher than those in the muscles and gills. Our results validate that one of the major sites of zinc-induced oxidative stress injury and repair in zebrafish is the intestine. Muscles and gills are inferior in comparison. However, the distribution of MDA levels in the intestine, muscles, and gills was not significantly different, demonstrating a strong repair capacity in the intestine. The intestine protects zebrafish from heavy metals through immunogenic and non-immunogenic mechanisms (Marinsek *et al.* 2022). High AChE activity in gills may be a reflection of more muscle activity.

Several heavy metals are known to target the microbiota in the digestive system (Bist & Choudhary 2022). The metabolic capacity and immune functions of the intestine can be reduced by heavy metal invasion, which is caused by disturbances in the intestinal microbial community and may further increase the probability of intestinal damage and disease (Duan *et al.* 2020). A significant increase in the Chao1 index of the zinc-exposed group indicated an increase in microbial community richness. A decrease in the Simpson index reflects the diversity of the intestinal microbiota. Consistent results have been reported (Wu *et al.* 2021). In the PCoA, clearly separated clusters were observed in the exposure group, demonstrating that chronic exposure to zinc altered the structure of the zebrafish intestinal microbiota.

In studies of the effect of zinc on the structure of the zebrafish intestinal microbiota, significant changes in the relative abundance of microorganisms at the phylum and genus levels were the most direct evidence of an effect of zinc. High-throughput sequencing analysis of 16S rRNA showed that the dominant intestinal microbiota at the fish phylum level was Fusobacteria, Tenericutes, Proteobacteria, Firmicutes, and Bacteroidetes. Similar findings can be retrieved from other studies (Dulski *et al.* 2020; Zhang *et al.* 2020). At the phylum level, zinc exposure increased Proteobacteria in the intestine, potentially contributing to a disruption in zebrafish intestinal microbiota (Tan *et al.* 2020). In addition, an increased relative abundance of Proteobacteria has been associated with the development of inflammatory bowel disease (Lobionda *et al.* 2019). Firmicutes and Bacteroides play vital roles in host lipid metabolism. Significant increases in Firmicutes and Bacteroides may lead to abnormal lipid metabolism in zebrafish (Wang *et al.* 2021). Bacteroides is an opportunistic pathogen that may lead to endogenous infections (Dong *et al.* 2020). For example, butyric acid produced by Fusobacterium can improve the inflammatory status of the intestinal mucosa and inhibit colon carcinoma (Zhang *et al.* 2020). Decreases in Fusobacterium and Tenericutes may herald increased odds of intestinal diseases in zebrafish.

Changes in the dominant microbiota at the genus level are another manifestation of the effects of zinc exposure on intestinal microbiota. The dominant bacteria in the control group were *Cetobacterium* and *Rhodobacter* spp. The dominant bacteria in the experimental groups were *Rhizobium* and *Enterococcus* spp. *Cetobacterium* is an anaerobic bacterium involved in vitamin B12 synthesis (Bai *et al.* 2019). *Cetobacterium* and *Rhodobacter* are more abundant in the intestine of healthy fish than that in diseased individuals (Li *et al.* 2017; Xue *et al.* 2017). A significant reduction in the relative abundance of bacteria in these two genera could potentially increase the chance of pathogenicity in fish. *Enterococcus* can improve antioxidant enzyme activity and disease resistance in fish while enhancing immunity (Kakade *et al.* 2020). The significant increase in the abundance of *Enterococcus* may be related to intestinal oxidative stress. As a probiotic in the intestine, the reason for the significant increase in *Rhizobium* is presumed that chronic exposure to zinc accelerates the production of its product coenzyme Q10, which plays a role in enhancing host immunity (Xia *et al.* 2018).

There are links and interactions between oxidative stress, neurotoxicity, and altered gut microbiota in zebrafish resistance to zinc stress. In this paper, a Pearson correlation analysis was performed between neurotoxicity biomarkers and oxidative stress parameters. AChE activity was significantly negatively correlated with SOD, GSH, and MT levels after chronic exposure to zinc in zebrafish ($p < 0.01$), suggesting that the neurotoxicity of zinc in zebrafish may be mediated by oxidative stress (Guo *et al.* 2022). In addition, there are two main perspectives on the interaction between the nervous system and the gut microbial community. First, neurotoxic substances can be metabolically detoxified by the gut microbiota directly after production through reduction and hydrolysis/defixation reactions (Claus *et al.* 2016). Second, certain gut microbes can produce short-chain fatty acids and acetylcholine, which can act on receptors in neurons and regulate the normal transmission of neural signals (Dempsey *et al.* 2019). The significant increase in the relative abundance of enterococci associated with oxidative stress at the genus level is also potential evidence that the oxidative system and gut microbes collaborate with each other in detoxification.

5. CONCLUSIONS

In summary, this study demonstrates the toxic effects of chronic zinc exposure on oxidative stress, neurotransmitters, and intestinal microbiota in adult zebrafish. The changes of MT, SOD, CAT, GSH, and MDA levels in intestines, muscles, and gills were analyzed. Differences in the oxidative stress response of different organs to zinc exposure were found. The intestine is a more important site of antioxidant defense than gills and muscles. Inhibition of AChE activity by zinc exposure may be related to oxidative stress, which adversely affects the nervous system of zebrafish. Furthermore, the accumulation of zinc in zebrafish induces ecological changes in the intestinal microbiota that are twofold. This is reflected in the increased abundance and reduced diversity of the intestinal microbiota. Zinc exposure not only resulted in a significant reduction of potentially beneficial bacteria and an increase in opportunistic pathogens but also induced an increase in microorganisms associated with oxidative stress and enhanced intestinal immunity.

ACKNOWLEDGEMENTS

This study was funded by the National Key R&D Program of China (No. 2022YFE0104900), the Natural Science Foundation of Hebei Province (No. D2021402035), the Opening Project of Key Laboratory of Road Traffic Environmental Protection Technology, Ministry of Transport, PRC, the Opening Project of Key Laboratory of Environment Controlled Aquaculture

(Dalian Ocean University) Ministry of Education (No. 202218), and the 2022 Graduate Research Capacity Enhancement Program of Beijing Technology and Business University.

AUTHOR CONTRIBUTIONS

Z.Y. administered the project, conducted investigation and funding acquisition, wrote the original draft, reviewed and edited the article. R.L. investigated the article, developed the methodology, and conducted formal analysis. S.L. investigated the article and performed visualization and funding acquisition. D.Q. performed visualization and supervised the work. G.L. performed visualization. C.W. administered the project, reviewed and edited the article, and performed funding acquisition. J.N. reviewed and edited the article. Y.S. administered the project and performed funding acquisition. H.H. reviewed and edited the article.

DATA AVAILABILITY STATEMENT

All relevant data are included in the paper or its Supplementary Information.

CONFLICT OF INTEREST

The authors declare there is no conflict.

REFERENCES

Arini, A., Gourves, P. Y., Gonzalez, P. & Baudrimont, M. 2015 Metal detoxification and gene expression regulation after a Cd and Zn contamination: an experimental study on *Danio rerio*. *Chemosphere* **128**, 125–133.

Avallone, B., Agnisola, C., Cerciello, R., Panzuto, R., Simoniello, P., Creti, P. & Motta, C. M. 2015 Structural and functional changes in the zebrafish (*Danio rerio*) skeletal muscle after cadmium exposure. *Cell Biol. Toxicol.* **31**, 273–283.

Bai, Z. A., Ren, T. J., Han, Y. Z., Rahman, M. M., Hu, Y. N., Li, Z. Q. & Jiang, Z. Q. 2019 Influences of dietary selenomethionine exposure on tissue accumulation, blood biochemical profiles, gene expression and intestinal microbiota of *Carassius auratus*. *Comp. Biochem. Physiol. C-Toxicol. Pharmacol.* **218**, 21–29.

Bist, P. & Choudhary, S. 2022 Impact of heavy metal toxicity on the gut microbiota and its relationship with metabolites and future probiotics strategy: a review. *Biol. Trace Elem. Res.* 34994948. https://doi.org/10.1007/s12011-021-03092-4 (online ahead of print).

Bokulich, N. A., Kaehler, B. D., Rideout, J. R., Dillon, M., Bolyen, E., Knight, R., Huttley, G. A. & Caporaso, J. G. 2018 Optimizing taxonomic classification of marker-gene amplicon sequences with QIIME 2's q2-feature-classifier plugin. *Microbiome* **6**, 90.

Capriello, T., Monteiro, S. M., Felix, L. M., Donizetti, A., Aliperti, V. & Ferrandino, I. 2021 Apoptosis, oxidative stress and genotoxicity in developing zebrafish after aluminium exposure. *Aquat. Toxicol.* **236**, 105872.

Castrillo, G., Teixeira, P. J., Paredes, S. H., Law, T. F., Lorenzo, L., Feltcher, M. E., Finkel, O. M., Breakfield, N. W., Mieczkowski, P., Jones, C. D., Paz-Ares, J. & Dangl, J. L. 2017 Root microbiota drive direct integration of phosphate stress and immunity. *Nature* **543**, 513–518.

Claus, S. P., Guillou, H. & Ellero-Simatos, S. 2016 The gut microbiota: a major player in the toxicity of environmental pollutants? *NPJ Biofilms Microbiomes* **2**, 16003.

Dane, H. & Sman, T. S. 2020 A morpho-histopathological study in the digestive tract of three fish species influenced with heavy metal pollution. *Chemosphere* **242**, 125212.

Dempsey, J. L., Little, M. & Cui, J. Y. 2019 Gut microbiome: an intermediary to neurotoxicity. *Neurotoxicology* **75**, 41–69.

Ding, J., Dong, Z., Chen, Q. L., Zheng, F., Wang, H. T. & Zhu, Y. G. 2019 Effects of long-term fertilization on the associated microbiota of soil collembolan. *Soil Biol. Biochem.* **130**, 141–149.

Dong, H. B., Sun, Y. X., Duan, Y. F., Li, H., Li, Y., Liu, Q. S., Wang, W. H. & Zhang, J. S. 2020 The effect of teprenone on the intestinal morphology and microbial community of Chinese sea bass (*Lateolabrax maculatus*) under intermittent hypoxic stress. *Fish Physiol. Biochem.* **46**, 2457.

Duan, H., Yu, L., Tian, F., Zhai, Q., Fan, L. & Chen, W. 2020 Gut microbiota: a target for heavy metal toxicity and a probiotic protective strategy. *Sci. Total Environ.* **742**, 140429.

Dulski, T., Kozlowski, K. & Ciesielski, S. 2020 Habitat and seasonality shape the structure of tench (*Tinca L.*) gut microbiome. *Sci. Rep.* **10**, 4460.

El-Kassas, H. Y. & Mohamed, L. A. 2014 Bioremediation of the textile waste effluent by *Chlorella vulgaris*. *Egypt J. Aquatic Res.* **40**, 301–308.

Gimenez, V. M. M., Bergam, I., Reiter, R. J. & Manucha, W. 2021 Metal ion homeostasis with emphasis on zinc and copper: potential crucial link to explain the non-classical antioxidative properties of vitamin D and melatonin. *Life Sci.* **281**, 119770.

Gozzard, E., Mayes, W. M., Potter, H. A. & Jarvis, A. P. 2011 Seasonal and spatial variation of diffuse (non-point) source zinc pollution in a historically metal mined river catchment, UK. *Environ. Pollut.* **159**, 3113–3122.

Guo, D., Luo, L. L., Kong, Y., Kuang, Z. Y., Wen, S. Y., Zhao, M., Zhang, W. G. & Fan, J. 2022 Enantioselective neurotoxicity and oxidative stress effects of paclobutrazol in zebrafish (*Danio rerio*). *Pestic. Biochem. Physiol.* **185**, 105136.

Horie, Y., Yonekura, K., Suzuki, A. & Takahashi, C. 2020 Zinc chloride influences embryonic development, growth, and Gh/Igf-1 gene expression during the early life stage in zebrafish (*Danio rerio*). *Comp. Biochem. Physiol. C-Toxicol. Pharmacol.* **230**, 108684.

Hu, W., Zhu, Q. L., Zheng, J. L. & Wen, Z. Y. 2022 Cadmium induced oxidative stress, endoplasmic reticulum (ER) stress and apoptosis with compensative responses towards the up-regulation of ribosome, protein processing in the ER, and protein export pathways in the liver of zebrafish. *Aquat. Toxicol.* **242**, 106023.

Hubner, C. & Haase, H. 2021 Interactions of zinc- and redox-signaling pathways. *Redox Biol.* **41**, 101916.

Jegatheesan, V., Pramanik, B. K., Chen, J., Navaratna, D., Chang, C. Y. & Shu, L. 2016 Treatment of textile wastewater with membrane bioreactor: a critical review. *Bioresour. Technol.* **204**, 202–212.

Jiang, N., Song, P., Li, X., Zhu, L., Wang, J., Yin, X. & Wang, J. 2022 Dibutyl phthalate induced oxidative stress and genotoxicity on adult zebrafish (*Danio rerio*) brain. *J. Hazard. Mater.* **424**, 127749.

Juncos, R., Arcagni, M., Squadrone, S., Rizzo, A., Arribere, M., Barriga, J. P., Battini, M. A., Campbell, L. M., Brizio, P., Abete, M. C. & Ribeiro Guevara, S. 2019 Interspecific differences in the bioaccumulation of arsenic of three Patagonian top predator fish: organ distribution and arsenic speciation. *Ecotoxicol. Environ. Saf.* **168**, 431–442.

Kakade, A., Salama, E. S., Pengya, F., Liu, P. & Li, X. K. 2020 Long-term exposure of high concentration heavy metals induced toxicity, fatality, and gut microbial dysbiosis in common carp, *Cyprinus carpio. Environ. Pollut.* **266**, 115293.

Kishor, R., Diane, P., Ganesh, D. S., Rijuta, G. S., Luiz, F. R. F., Muhammad, B. R. C. & Ram, N. B. 2021 Ecotoxicological and health concerns of persistent coloring pollutants of textile industry wastewater and treatment approaches for environmental safety. *J. Environ. Chem. Eng.* **9**, 2.

Lee, S. R. 2018 Critical role of zinc as either an antioxidant or a prooxidant in cellular systems. *Oxid. Med. Cell. Longevity* **2018**, 9156285.

Li, T., Li, H., Gatesoupe, F. J., She, R., Lin, Q., Yan, X., Li, J. & Li, X. 2017 Bacterial signatures of 'red-operculum' disease in the gut of crucian carp (*Carassius auratus*). *Microb. Ecol.* **74**, 510–521.

Lobionda, S., Sittipo, P., Kwon, H. Y. & Lee, Y. K. 2019 The role of gut microbiota in intestinal inflammation with respect to diet and extrinsic stressors. *Microorganisms* **7**, 8.

Luzio, A., Parra, S., Costa, B., Santos, D., Alvaro, A. R. & Monteiro, S. M. 2021 Copper impair autophagy on zebrafish (*Danio rerio*) gill epithelium. *Environ. Toxicol. Pharmacol.* **86**, 103674.

Maret, W. 2006 Zinc coordination environments in proteins as redox sensors and signal transducers. *Antioxid. Redox Signal.* **8**, 9–10.

Maret, W. 2019 The redox biology of redox-inert zinc ions. *Free Radic. Biol. Med.* **134**, 311–326.

Marinsek, G. P., Choueri, P. K. G., Choueri, R. B., Abessa, D. M. D., Goncalves, A. R. N., Bortolotto, L. B. & Mari, R. D. 2022 Integrated analysis of fish intestine biomarkers: complementary tools for pollution assessment. *Mar. Pollut. Bull.* **178**, 113590.

McRae, N. K., Gaw, S. & Glover, C. N. 2016 Mechanisms of zinc toxicity in the galaxiid fish, Galaxias maculatus. *Comp. Biochem. Physiol. C-Toxicol. Pharmacol.* **179**, 184–190.

Muthulakshmi, S., Maharajan, K., Habibi, H. R., Kadirvelu, K. & Venkataramana, M. 2018 Zearalenone induced embryo and neurotoxicity in zebrafish model (*Danio rerio*): role of oxidative stress revealed by a multi biomarker study. *Chemosphere* **198**, 111–121.

Nidheesh, P. V., Divyapriya, G., Titchou, F. E. & Hamdani, M. 2022 Treatment of textile wastewater by sulfate radical based advanced oxidation processes. *Sep. Purif. Technol.* **293**, 121115.

Parlak, V. 2018 Evaluation of apoptosis, oxidative stress responses, AChE activity and body malformations in zebrafish (*Danio rerio*) embryos exposed to deltamethrin. *Chemosphere* **207**, 397–403.

Puar, P., Naderi, M., Niyogi, S. & Kwong, R. W. M. 2021 Using zebrafish as a model to assess the individual and combined effects of sub-lethal waterborne and dietary zinc exposure during development. *Environ. Pollut.* **284**, 117377.

Pullaguri, N., Nema, S., Bhargava, Y. & Bhargava, A. 2020 Triclosan alters adult zebrafish behavior and targets acetylcholinesterase activity and expression. *Environ. Toxicol. Pharmacol.* **75**, 103311.

Saiki, P., Mello-Andrade, F., Gomes, T. & Rocha, T. L. 2021 Sediment toxicity assessment using zebrafish (*Danio rerio*) as a model system: historical review, research gaps and trends. *Sci. Total Environ.* **793**, 148633.

Santos, D., Luzio, A., Felix, L., Bellas, J. & Monteiro, S. M. 2022 Oxidative stress, apoptosis and serotonergic system changes in zebrafish (*Danio rerio*) gills after long-term exposure to microplastics and copper. *Comp. Biochem. Physiol. C-Toxicol. Pharmacol.* **258**, 109363.

Shahjahan, M., Taslima, K., Rahman, M. S., Al-Emran, M., Alam, S. I. & Faggio, C. 2022 Effects of heavy metals on fish physiology – a review. *Chemosphere* **300**, 134519.

Sharma, R. & Jindal, R. 2020 Assessment of cypermethrin induced hepatic toxicity in *Catla*: a multiple biomarker approach. *Environ. Res.* **184**, 109359.

Sharma, K., Sharma, P., Dhiman, S. K., Chadha, P. & Saini, H. S. 2022 Biochemical, genotoxic, histological and ultrastructural effects on liver and gills of freshwater fish *Channa punctatus* exposed to textile industry intermediate 2 ABS. *Chemosphere* **287**, 132103.

Si, L. F., Wang, C. C., Guo, S. N., Zheng, J. L. & Xia, H. 2019 The lagged effects of environmentally relevant zinc on non-specific immunity in zebrafish. *Chemosphere* **214**, 85–93.

Skidmore, J. F. 1970 Respiration and osmoregulation in rainbow trout with gills damaged by zinc sulphate. *J. Exp. Biol.* **52**, 481–494.

Sun, H. J., Zhang, Y., Zhang, J. Y., Lin, H., Chen, J. & Hong, H. 2019 The toxicity of 2,6-dichlorobenzoquinone on the early life stage of zebrafish: a survey on the endpoints at developmental toxicity, oxidative stress, genotoxicity and cytotoxicity. *Environ. Pollut.* **245**, 719–724.

Sun, H. J., Zhao, W. J., Teng, X. Q., Shu, S. P., Li, S. W., Hong, H. C. & Guan, D. X. 2020 Antioxidant responses and pathological changes in the gill of zebrafish (*Danio rerio*) after chronic exposure to arsenite at its reference dose. *Ecotoxicol. Environ. Saf.* **200**, 110743.

Tan, P., Wu, X., Zhu, W. L., Lou, B., Chen, R. Y. & Wang, L. G. 2020 Effect of tributyrin supplementation in high-soya bean meal diet on growth performance, body composition, intestine morphology and microbiota of juvenile yellow drum (*Nibea albiflora*). *Aquac. Res.* **51**, 2004–2019.

Tao, Y., Li, Z., Yang, Y., Jiao, Y., Qu, J., Wang, Y. & Zhang, Y. 2022 Effects of common environmental endocrine-disrupting chemicals on zebrafish behavior. *Water Res.* **208**, 117826.

Tsai, J. W. & Liao, C. M. 2006 A dose-based modeling approach for accumulation and toxicity of arsenic in tilapia *Oreochromis mossambicus*. *Environ. Toxicol.* **21**, 8–21.

Wang, N., Jiang, M., Zhang, P., Shu, H., Li, Y., Guo, Z. & Li, Y. 2020 Amelioration of Cd-induced bioaccumulation, oxidative stress and intestinal microbiota by *Bacillus cereus* in *Carassius auratus gibelio*. *Chemosphere* **245**, 125613.

Wang, W. Z., Huang, J. S., Zhang, J. D., Wang, Z. L., Li, H. J., Amenyogbe, E. & Chen, G. 2021 Effects of hypoxia stress on the intestinal microflora of juvenile of cobia (*Rachycentron canadum*). *Aquaculture* **536**, 736419.

Wu, Y., Huang, J., Deng, M., Jin, Y., Yang, H., Liu, Y., Cao, Q., Mennigen, J. A. & Tu, W. 2019 Acute exposure to environmentally relevant concentrations of Chinese PFOS alternative F-53B induces oxidative stress in early developing zebrafish. *Chemosphere* **235**, 945–951.

Wu, L., Xu, Y., Lv, X., Chang, X., Ma, X., Tian, X., Shi, X., Li, X. & Kong, X. 2021 Impacts of an azo food dye tartrazine uptake on intestinal barrier, oxidative stress, inflammatory response and intestinal microbiome in crucian carp (*Carassius auratus*). *Ecotoxicol. Environ. Saf.* **223**, 112551.

Xavier, N. D. D., Nandan, S. B., Jayachandran, P. R., Anu, P. R., Midhun, A. M. & Mohan, D. 2019 Chronic effects of copper and zinc on the fish, *Etroplus suratensis* (Bloch, 1790) by continuous flow through (CFT) bioassay. *Mar. Environ. Res.* **143**, 141–157.

Xia, Y., Lu, M., Chen, G., Cao, J., Gao, F., Wang, M., Liu, Z., Zhang, D., Zhu, H. & Yi, M. 2018 Effects of dietary *Lactobacillus rhamnosus* JCM1136 and *Lactococcus lactis* subsp. lactis JCM5805 on the growth, intestinal microbiota, morphology, immune response and disease resistance of juvenile Nile tilapia, *Oreochromis niloticus*. *Fish Shellfish Immunol.* **76**, 368–379.

Xu, Z., Wang, P., Wang, H., Yu, Z. H., Au-Yeung, H. Y., Hirayama, T., Sun, H. & Yan, A. 2019 Zinc excess increases cellular demand for iron and decreases tolerance to copper in *Escherichia coli*. *J. Biol. Chem.* **294**, 16978–16991.

Xue, S., Xu, W., Wei, J. & Sun, J. 2017 Impact of environmental bacterial communities on fish health in marine recirculating aquaculture systems. *Vet. Microbiol.* **203**, 34–39.

Zeng, L., Ai, C. X., Zhang, J. S. & Pan, Y. 2019 Toxicological effects of waterborne Zn on the proximal and distal intestines of large yellow croaker *Larimichthys crocea*. *Ecotoxicol. Environ. Saf.* **174**, 324–333.

Zhang, Y., Li, Z. Y., Kholodkevich, S., Sharov, A., Chen, C., Feng, Y. J., Ren, N. Q. & Sun, K. 2020 Effects of cadmium on intestinal histology and microbiota in freshwater crayfish (*Procambarus clarkii*). *Chemosphere* **242**, 125105.

Zhang, J., Meng, H., Kong, X., Cheng, X., Ma, T., He, H., Du, W., Yang, S., Li, S. & Zhang, L. 2021 Combined effects of polyethylene and organic contaminant on zebrafish (*Danio rerio*): accumulation of 9-Nitroanthracene, biomarkers and intestinal microbiota. *Environ. Pollut.* **277**, 116767.

Zhang, J., Tan, Q. G., Huang, L., Ye, Z., Wang, X., Xiao, T., Wu, Y., Zhang, W. & Yan, B. 2022 Intestinal uptake and low transformation increase the bioaccumulation of inorganic arsenic in freshwater zebrafish. *J. Hazard. Mater.* **434**, 128904.

Zheng, J. L., Zhu, Q. L., Hu, X. C., Parsons, D., Lawson, R. & Hogstrand, C. 2022 Transgenerational effects of zinc in zebrafish following early life stage exposure. *Sci. Total Environ.* **828**, 154443.

First received 8 November 2022; accepted in revised form 29 December 2022. Available online 23 January 2023

IWA Publishing's authorised EU representative for General Product
Safety Regulations is Diane D'Arras, 15 rue Duret, 75116 Paris,
France, e-mail: safety@iwap.co.uk.

Printed and bound by CPI Group (UK) Ltd, Croydon, CR0 4YY

12/05/2026

02108892-0001